Analyzing Emotion in Spontaneous Speech

Rupayan Chakraborty · Meghna Pandharipande
Sunil Kumar Kopparapu

Analyzing Emotion
in Spontaneous Speech

Rupayan Chakraborty
Speech and Natural Language Processing
TCS Research and Innovation—Mumbai
Tata Consultancy Services Limited
Thane, Maharashtra
India

Sunil Kumar Kopparapu
Speech and Natural Language Processing
TCS Research and Innovation—Mumbai
Tata Consultancy Services Limited
Thane, Maharashtra
India

Meghna Pandharipande
Speech and Natural Language Processing
TCS Research and Innovation—Mumbai
Tata Consultancy Services Limited
Thane, Maharashtra
India

ISBN 978-981-13-5668-1   ISBN 978-981-10-7674-9   (eBook)
https://doi.org/10.1007/978-981-10-7674-9

Printed on acid-free paper

This Springer imprint is published by Springer Nature
The registered company is Springer Nature Singapore Pte Ltd.
The registered company address is: 152 Beach Road, #21-01/04 Gateway East, Singapore 189721, Singapore

I dedicate this book to my Parents, my Sister, and my Wife

—Rupayan Chakraborty

This book is for you, Baba. You left fingerprints of grace on our lives. As you look down from heaven, I hope you're proud of your little girl.

—Meghna Pandharipande

To all those who do not express their emotions explicitly and hope to be understood!

—Sunil Kumar Kopparapu

Preface

Emotions are part and parcel of our day-to-day lives. Sometimes we express them, and sometimes we do not express them explicitly. Though they are interlaced in our day-to-day lives, it is by no means easy to define emotion, and as a result, there is no consensus on an acceptable definition of emotion. Wikipedia defines it as "any conscious experience characterized by intense mental activity and a high degree of pleasure or displeasure." Emotion can be loosely defined as a *feeling* that a normal human experiences based on their circumstances and can be influenced by other animate and inanimate objects as well as themselves. Some people might display their emotions explicitly because of the circumstances in many ways; for example, anger could be demonstrated by destroying an object (action) or by facial expression (visual) or by speaking in a certain way (audio). It is also possible that some people might not demonstrate their emotions because of their situation or because of their social settings or because of their personality. But the fact remains that there are traces of one emotion or the other which are part of our being all the time as we go about our lives.

In this monograph, we look at emotions embedded in audio, especially when they occur in our day-to-day interactions in a natural setting called spontaneous spoken conversation. While it is not very difficult to analyze speech and recognize emotions of a customer who calls a service provider in anger and wants to complain about a poor service, in reality most often the emotions of the customer are either hidden or not demonstrated loudly or explicitly. It is in this realistic scenario that automatically identifying the emotion of the customer helps the service provider in upselling or providing a service to suit the emotional state of the customer. The overall effect of being able to identify the emotion of the speaker helps in providing a better user experience as well as retaining the customer.

This monograph tries to uncover aspects that enable identification of emotions, which are not explicitly demonstrated by the speaker; these subtle emotions generally occur in spontaneous speech (non-acted). Most of the work reported here is the outcome of the work that we have been pursuing, and a lot of what we say here has appeared in some form or the other in the literature published by us.

In brief, the monograph is divided into six chapters which focus on analyzing emotion in spontaneous speech. The monograph details the kind of information that is embedded in speech and then highlights the difficulties and challenges that exist in machine recognizing emotions in spontaneous speech. A brief literature survey brings out the existence of the most used emotion speech databases. A framework is proposed which makes use of knowledge associated with the speech to enable automatic emotion recognition for spontaneous speech. The monograph also dwells on several ways and methods of improving emotion recognition in spontaneous speech. And finally, a couple of case studies which are specifically applicable to voice-based call centers are presented.

In Chap. 1, we describe the different dimensions of information that are embedded in a speech signal in general and briefly mention the use of each of them in a highly digitized world. We then concentrate on "The How" dimension of the speech signal which refers to the emotion expressed in the speech signal and elaborate on the need for emotion analysis. We move to emotions in spontaneous speech and bring out the several challenges facing automatic recognition of emotion in spontaneous speech. This forms the basis for the rest of the monograph.

In Chap. 2, we review the work in the area of audio emotion recognition. While most of the reported work is in the area of acted speech, there has been some work reported in the area of spontaneous speech. We devote a complete section on the patent literature to bring out the industrial applicability of audio emotion recognition. We also review the databases that are most often used in the literature and are available to score the performance of any new algorithm that one might come up to address automatic audio emotion recognition. We also briefly mention a mechanism to build a database suitable for emotion recognition experiments.

In Chap. 3, we describe a framework that we built over a period of time to enable robust recognition of emotions in spontaneous speech. The proposed emotion recognition framework is knowledge-driven and hence scalable in the sense that more knowledge blocks can be appended to the framework, additionally, the framework is able to address both acted and spontaneous speech emotion recognition.

In Chap. 4, we capture methods of improving the performance of an automatic audio emotion recognition system in terms of (a) segmenting the audio into smaller entities, (b) choice of robust features, and (c) use of error-correcting codes to improve the classification accuracies.

In Chap. 5, we discuss two case studies. The first one deals with a method to mine call center calls with similar emotions in a large call center conversation audio repository, and the second one deals with identifying the *felt* emotion of a person after having watched a small movie clip mostly capturing our participation in MediaEval workshop. We conclude in Chap. 6.

Thane (West), Maharastra, India   Rupayan Chakraborty
(Loc: 72.977265, 19.225129)   Meghna Pandharipande
December 2017   Sunil Kumar Kopparapu

Acknowledgements

Several people in very many different ways have contributed in the journey of this monograph. We would like to acknowledge all of them with utmost sincerity, as they say, from the bottom of our hearts.

The work reported here is based on discussions and interactions with the team at TCS Research and Innovation—Mumbai. We have liberally made use of the write-ups, documents, presentations that we conceptualized and built together over the last couple of years as part of our ongoing work in the area of audio emotion recognition. Our discussion with customers and clients during the course of this work has influenced some of the content; we gratefully acknowledge their role in shaping the material. We would like to thank the SERES team and in particular Chitralekha and Mithun for making available the IVR-SERES data.

We would also like to thank folks from Springer for following up and making sure we had the prose ready and in time. But for them, we might just not have found the energy to put together all this material into a monograph.

In today's digital world, it is but imperative that we make use of several pieces of software to bind together material. We acknowledge LaTex, xfig, Dia, freeplane, OpenOffice, and the folks who have developed these wonderful pieces of software.

Lastly, we take complete responsibility for any deficiency in this monograph, be it typo, grammar, or content. To err is human!

Contents

About the Authors

Rupayan Chakraborty (Member IEEE) works as a Scientist at TCS Research and Innovation—Mumbai. He has been working in the area of speech and audio signal processing since 2008 and was involved in academic research prior to joining TCS. He worked as a Researcher at the Computer Vision and Pattern Recognition (CVPR) Unit of the Indian Statistical Institute (ISI), Kolkata. He obtained Ph.D. degree from TALP Research Centre, UPC, Barcelona, Spain, in December 2013, for his work in the area of computational acoustic scene analysis, while working on the project "Speech and Audio Recognition for Ambient Intelligence (SARAI)". After completing his Ph.D., he was a Visiting Scientist at the CVPR Unit of ISI, Kolkata, for 1 year. He has published research works in top-tier conferences and journals. He is currently working in the area of "speech emotion recognition and analysis."

Meghna Pandharipande received her Bachelor of Engineering (BE) in Electronics and Telecommunication in June 2002 from Amravati University, Amravati. Between September 2002 and December 2003, she was a Faculty Member of the Department of Electronics and Telecommunication at Shah and Anchor Kutchhi Engineering College, Mumbai. In 2004, she completed her certification in Embedded Systems at CMC, Mumbai, and then worked as a Lotus Notes Developer in a startup ATS, Mumbai, for a year. Since June 2005, she has been with TCS (having first joined the Cognitive Systems Research Laboratory, Tata InfoTech Limited, under Prof. P. V. S. Rao), and since 2006, she has been working as a Researcher at TCS Research and Innovation—Mumbai. Her research interest is in the area of speech signal processing and has been working extensively on building systems that can process all aspects of spoken speech. More recently, she has been researching nonlinguistic aspects of speech processing, like speaking rate and emotion detection from speech.

Sunil Kumar Kopparapu (Senior Member, IEEE; ACM Senior Member, India) obtained his doctoral degree in Electrical Engineering from the Indian Institute of Technology Bombay, Mumbai, India, in 1997.

Between 1997 and 2000, he was with the Automation Group, Commonwealth Scientific and Industrial Research Organization (CSIRO), Brisbane, Australia, working on practical image processing and 3D vision problems, mainly for the benefit of the Australian mining industry.

Prior to joining the Cognitive Systems Research Laboratory (CSRL), Tata InfoTech Limited, as a Senior Research Member in 2001, he was an expert for developing virtual self-line of e-commerce products for the R&D Group at Aquila Technologies Private Limited, India.

In his current role as a Principal Scientist at TCS Research and Innovation—Mumbai, he is actively working in the areas of speech, script, image, and natural language processing with a focus on building usable systems for mass use in Indian conditions. He has co-authored a book titled "Bayesian Approach to Image Interpretation" and more recently a Springer Brief on Non-linguistic Analysis of Call Center Conversations.

Chapter 1
Introduction

1.1 Information Embedded in Speech

Speech is and has been the most used natural mode of communication between humans across geographies. Speech is interlaced with many intended meanings depending on the social context and cultural settings. Most importantly, any spoken utterance has primarily three important components associated with it, namely nonlinguistic, linguistic, and paralinguistic which can also be categorized as "The Who," "The What," and "The How" dimensions (see Fig. 1.1), respectively.

Assume a spoken utterance /*In the midst of darkness light exists*/. The nonlinguistic aspect of analysis would be to identify (a) who spoke it or associating a name to the speaker, (b) what was the gender of the speaker, (c) what was the emotional state of the speaker, (d) in what language did the speaker speak, etc., while from the linguistic angle it would be (e) what was spoken in this case the text corresponding to the utterance, namely "In the midst of darkness light exists". Most of the ongoing speech research revolves around extracting or analyzing one or more of these three aspects of speech.

An audio signal, as seen in Fig. 1.1, primarily has three dimensions associated with it, namely "The Who," "The What," and "The How". "The Who" dimension looks at the person behind the voice, while "The What" dimension looks at the linguistic content in the voice, and "The How" dimension looks at the expressed emotion in the voice. The availability of large volumes of speech data and the rapid advances in machine learning have made the machines of today capable of "understanding" what ("The What") is being spoken (speech to text or automatic speech recognition) to a certain extent, who ("The Who") is speaking (speech biometric) quite reliably in a constrained scenario, and ("The How") how (emotional state) of the voice to a lesser extent. All the three dimensions of speech are being explored world over. While each of these dimensions has been sufficiently explored, they come with their own set of challenges and along with them are great opportunities that are begging to be explored. In this monograph we concentrate on "The How" aspect of a spoken utterance, namely emotion recognition with an extra emphasis on spontaneous

© Springer Nature Singapore Pte Ltd. 2017
R. Chakraborty et al., *Analyzing Emotion in Spontaneous Speech*,
https://doi.org/10.1007/978-981-10-7674-9_1

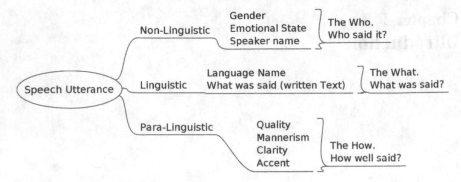

Fig. 1.1 Information embedded in a speech signal

speech. One of the aims of this monograph is to give the reader a feel for the existing challenges in recognizing emotions in spontaneous spoken speech in addition to sharing our own experiences, working with realistic data, in trying to overcome these challenges.

1.2   Embedded Emotion in Speech

Though there is no one agreed definition of emotion, historically, as seen in Fig. 1.2, the two well-articulated affective dimensions, namely arousal (also referred to as activation or intensity, along the x-axis) and valence (also referred to as attractiveness or goodness for positive valence, along the y-axis), have been used to represent emotion. Any point in this two-dimensional space can be looked upon as a vector which can be mapped onto an emotion label (e.g., happy, anger, sad). Table 1.1 gives the mapping of a few emotions in terms of the affective states. For example, +ve arousal and +ve valence (first quadrant, represented by $(++, ++)$)

Fig. 1.2 Sample emotions in the (arousal, valence) affective 2D space

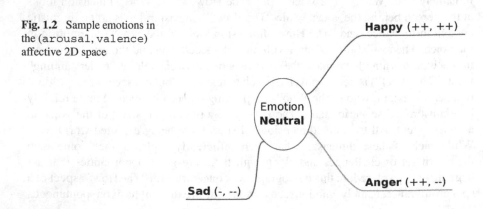

Table 1.1 Emotion labels mapped to the two affective states (arousal, valence). "+ +" represents greater value compared to "+"

Affective state		Emotion
arousal	valence	Label
++	++	happy
++	−−	anger
−	+	neutral
−	−−	sad

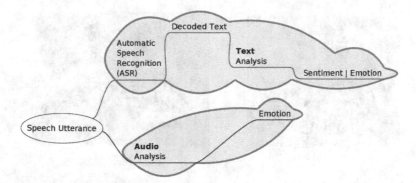

Fig. 1.3 Determining emotion from speech in one and two steps

would represent happy while +ve arousal and −ve valence could represent anger (++, −−) and so on.

The ability to perceiving emotions, expressed or otherwise, from different human-generated real-life cues (signals) is an inherent characteristic of a human. For this reason, the ability to detect or identify emotions plays a very significant role in an intelligent human–computer interaction system. Machine perception of human emotion not only helps machine to communicate more humanely (as seen in Sect. 1.3), but it also helps in improving the performance of other associated technologies like automatic speech recognition (ASR) and speaker identification (SI) which are an integral part of the human–machine interaction systems like voice-enabled self-help systems.

Identification or recognition of emotion from an audio speech utterance can be performed in one of the two ways as shown in Fig. 1.3. The most common and conventional approach adopted is the two-step approach (see Fig. 1.4). In the two-step approach, initially the speech utterance is converted into text using an automatic speech recognition system. This is followed by the second step of analyzing the text output for sentiments or emotion. The sentiment extracted from the linguistic text is then attributed to the spoken audio utterance.

There are several challenges in this two-step approach. While it goes without saying that there is a need to a priori know the language in which the audio utterance

Step 1:	Speech to text using automatic speech recognition (ASR)
Step 2:	Analyzing linguistic (text) content

- Based on *what* is spoken
- Extraction of text from audio
- Learning | Classification (text)

Pros | Cons: Language dependent; dependent on performance of ASR

Fig. 1.4 Speech → text → emotion (2 steps)

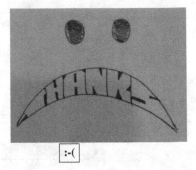

Fig. 1.5 Linguistic words (e.g., thanks) need not translate to expressed emotions

was spoken so that an appropriate language-specific ASR could be used to convert the utterance to text, more importantly there is a need for a *robust* ASR to support the speech to text conversion process with greater degree of accuracy. Finally, there is the need also for a text analysis system to determine the sentiment from the decoded words. Clearly, this approach is dependent on the linguistic words ("The What") that were spoken, which implies the systems that adopt the two-step approach are unable to distinguish between a sarcastic */thanks/* and a genuine */thanks/* (see Fig. 1.5). The emotion recognition process is based on the learning of a classifier based on the textual information. Clearly the two-step approach is language dependent, and the accuracy of the system directly depends on the performing accuracy of the ASR used and the linguistic words uttered in the audio.

On the other hand (see Fig. 1.6), the one-step approach directly operates on the spoken utterance and recognizes the emotion from the extracted audio features. This approach is only dependent on how the ("The How") speech was uttered and not necessarily based on what was spoken, thus making this approach able to distinguish a genuine */thanks/* from a sarcastic */thanks/*. This approach makes it largely independent of the spoken language, and subsequently it is also able to handle utterances with mixed language. Note that use of mixed language is very dominant in day-to-day natural spoken conversions. In this monograph we will emphasize on the second approach, namely the one-step ("The How") approach.

Step 1:   Analyzing non-linguistic (audio) content

- Based on *how* spoken
- Extraction of features from audio
- Learning | Classification (audio features)

Pros | Cons: No need of an ASR. Can handle mixed language utterance.

Fig. 1.6 Speech → emotion (1 step)

Emotions when explicitly expressed by a speaker (monologues or dialogues of an actors for example) are easy for a machine to recognize. However, recognizing emotion in audio generated during day-to-day interaction between people, which is predominantly naturally spoken spontaneous speech, is not easy for machines to identify the emotion even though emotion of the speaker is implicitly embedded in their spoken utterances. There are several reasons for this, but the most important one is because people, especially non-actors, do not *strongly* express their emotion when they speak, thus making it difficult for machines to recognize the emotion (or distinguish one emotion from another) embedded in their speech.

1.3 Need for Emotion Analysis

With the mushrooming of services industry there has been a significant growth in the voice-based (a) call centers (VbCC) and (b) self-help systems where identifying emotion in customer speech is crucial [18] for the business of an enterprise. The primary goal of a VbCC is to maintain a high level of customer satisfaction index (CSI) which means understanding the customer continuously during the interaction in real time and automatically. This *just in time* understanding of the customer helps the enterprise make a decision on how to communicate (what to say, how to say) with the customer. While several attributes related to the customer are *a priori* available, thanks to the customer profiling achieved due to the advances in data mining, the one crucial attribute that is missing is the emotion of the customer at that instance of conversation. The real-time identification of the emotion of the customer can help the agent plan what and how to converse to keep the customer happy and additionally allows the agent to sense when to make an up-sell successful pitch.

The initial research work in the area of audio emotion recognition has been fairly well demonstrated and validated on acted speech (e.g., [3, 63, 95, 114]) corpus. Emotions in audio expressed by *trained* actors are much easier to recognize than spontaneous speech, primarily because the acted emotions are explicitly and dramatically magnified by them. This, on purpose magnified, expression helps in distinguishing one emotion from another. However, when the expression of the emotion is neither magnified, explicit, or loud, it is very difficult to distinguish one emotion of the speaker from another. The mild, undemonstrated emotion is most likely to occur in spontaneous natural day-to-day conversational speech, thus making recognition of emotion in spontaneous natural speech very challenging.

Assume a voice-based self-help system operated by a car insurance company. Let us consider the scenario of a customer calling the self-help system to claim insurance because of an accident.

AGENT :-) :/*Thanks you for calling xyz Insurance, how may I help you?/*

CUSTOMER :-(:/*I just had an accident and need to claim insurance and tow the car urgently please/*

Note that the AGENT is chirpy :-) when the CUSTOMER calls in, and the reaction of the self-help agent (AGENT) after listening to the customer could be any of the following:

R_1 :-) :/*Can I have your* 18 *digit insurance policy number/*

R_2 :-) :/*Sorry to hear this. Where is your vehicle currently/*

R_3 :-(:/*Sorry to hear this. Where is your vehicle currently/*

Clearly, R_1 is the response of the most self-help system today which will in all probability result in a poor customer experience both in terms of the spoken content and the way :-) it is spoken. A speech understanding system probably would respond as R_2 wherein it would express empathy /*Sorry to hear this./* while continuing to be chirpy :-) . While R_3 which not only understands what the CUSTOMER is saying but is able to empathize with the CUSTOMER communicate back with an emotion :-(in tune with the emotion of the CUSTOMER. Clearly, R_3 kind of response will need to recognize the emotional state of the customer. So detecting emotions in spontaneous conversations can lead to better user experience (Ux).

1.4 Spontaneous Speech Emotion Analysis

There has been some serious work in the area of emotion recognition in general and emotion from audio in particular; however a large portion of this work has been evaluated on acted speech, and not much work has been done on spontaneous speech. A typical day-to-day conversation between people could be considered as an example of spontaneous speech, the kind of conversational language that does not adhere to conventional language grammar. The increasing level of difficulty for a machine to

Table 1.2 Degree of challenges in audio emotion recognition

Difficulty level	Less challenging $\longrightarrow$	Very challenging
Type	Acted $\longrightarrow$	Spontaneous
Media	Video $\longrightarrow$	Only audio
	Read speech $\longrightarrow$	Natural speech
Language	Single language $\longrightarrow$	Mixed language
Clarity	Clean speech $\longrightarrow$	Noisy speech
Distance	Close talk $\longrightarrow$	Far field
Corpus	Available $\longrightarrow$	Not available
Annotation	Good $\longrightarrow$	Poor
Context	Available $\longrightarrow$	Not available
Good ASR	Available $\longrightarrow$	Not available

recognize emotion is captured in Table 1.2. We now elaborate in the next section the reasons that make emotion recognition of naturally spoken spontaneous speech utterances a challenge.

1.4.1 Challenges

1. **Intensity of Emotion in Spontaneous Speech**

   Usually spontaneous speech lacks intensity unlike acted speech which exhibits higher degree of intensity, both in the `arousal` and the `valence` dimensions. The greater intensity results in a larger magnitude emotion vector compared to the spontaneous (non-acted) speech. Subsequently, as can be observed in Fig. 1.7, the distance between the vectors representing two different emotions for spontaneous speech is small. This makes it hard to distinguish one emotion from the other. On the other hand, the magnitude of the vectors for acted speech is larger for acted speech, making it easier to discriminate one emotion from another.

   To demonstrate this observation, let the first quadrant $(+, +)$ in Fig. 1.7 represent an emotion E_1 and let the fourth quadrant $(+, -)$ represent an emotion E_2 such that $E_1 \neq E_2$. Notice that Δr is the distance between E_1 and E_2 for spontaneous speech, while Δt is the distance between the same emotions E_1 and E_2 for acted speech. Clearly, $\Delta t > \Delta r$ which implies that emotions E_1 and E_2 are more separable when they represent acted speech compared to when they represent spontaneous natural speech. Or in other words it requires higher classification error (Δt) to misrecognize emotion E_1 as emotion E_2 and vice-versa for acted speech compared to the spontaneous speech (Δt). This is one of the reasons which makes spontaneous speech emotion recognition a challenging problem.

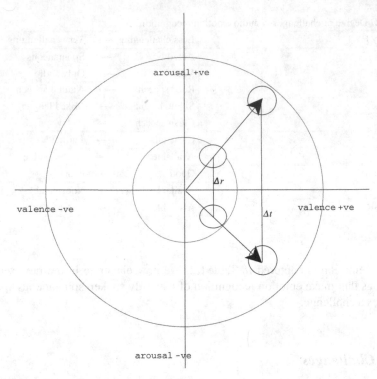

Fig. 1.7 Emotions in the two-dimensional affective (`arousal`, `valence`) space. Emotion in spontaneous speech is very subtle compared to the acted speech

2. **Data Mismatch**
   The speech data associated with acted and spontaneous speech differ. As a result, the techniques and algorithms that work for acted speech do not necessarily work for spontaneous speech. Subsequently, the rich literature which addresses acted audio emotion recognition cannot be used for spontaneous audio emotion recognition especially because most of the work in emotion literature is based on machine learning. The working of any machine learning-based system is based on training (e.g., support vector machine, artificial neural networks, deep neural networks) on a large set of emotion (or affective state)-labeled speech data. The trained classifier is then used to recognize the test dataset. Speech emotion recognition systems that perform with high accuracies on acted speech datasets do not perform as well on spontaneous and realistic natural speech [98] because of the mismatch between the train (acted) and test (spontaneous) datasets. This is yet another challenge in addressing the problem of spontaneous speech emotion recognition.
   Note that the challenge arising out of data mismatch can be overcome if a good emotion-labeled spontaneous speech corpus was available so that the corpus could be used to train a machine learning system for spontaneous speech.

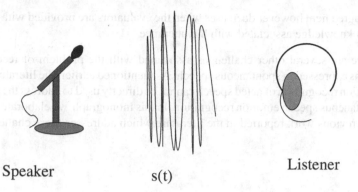

Speaker s(t) Listener

Fig. 1.8 Decoded versus encoded emotions

3. **Spontaneous Speech Corpus**

A spoken utterance has generally two point of views: One forms the point of view of the speaker and the another the point of view of the listener (see Fig. 1.8). When the speaker annotates his own utterance $s(t)$ in terms of emotion labels it is called encoded emotion and when the listener, who is different from the speaker, listens to $s(t)$ and then annotates $s(t)$ with an emotion label is called decoded emotion (see Appendix B for more detailed discussion).

While one can expect both decoded and encoded emotion labels to be the same for acted speech, the same, however, is not true for spontaneous speech. The expressed emotion not being as strong in spontaneous speech results in a wider gap between the encoded and decoded emotion labels. One can guess that if the speaker correctly labels the utterance, the encoded emotion label is closer to the actual emotion expressed in the audio.

Building a realistic spontaneous speech corpus would require encoded emotion labels, namely there is a need for a person to first speak when she is in a certain emotional state in addition to correctly annotating what she spoke; generating such realistic spontaneous data corpus is extremely difficult and is a huge challenge (see Sect. 2.4).

4. **Emotion Labeling Spontaneous Speech**

Difficulty in building encoded emotion-labeled speech corpus means one has to settle for the next best thing, namely have a decoded emotion-labeled spontaneous speech corpus. However, the problem associated with emotion recognition of spontaneous speech is the availability of a reliably annotated spontaneous speech corpus suitable for emotion recognition. The inability to annotate spontaneous speech corpus is basically because of the lower degree of emotion expression (as seen in Fig. 1.7) which results in the decoded emotion being far from the actual emotion embedded in the utterance.

As shown in [15] there is large degree of disagreement among the annotators when they are asked to annotate spontaneous spoken utterances. The disagreement however decreases when the evaluators are provided with the context knowledge associated with the utterance.

There are several other challenges associated with the problem of recognizing emotions expressed in spontaneous speech. As mentioned earlier, the literature based on emotion recognition of acted speech cannot be directly used to address the problem of spontaneous speech emotion recognition. In this monograph we elaborate on some of our previous work reported in the literature which addresses these challenges.

Chapter 2
Literature Survey

2.1 Brief Survey of Spontaneous Speech Emotion Recognition

Explicitly demonstrated emotions in speech, especially by actors are easy for a machine to recognize compared to spontaneous conversations that occur in human–human (day-to-day natural conversation) or human–machine (example call center conversations) interactions as discussed in the previous chapter. One of the main reasons for this is that people, especially non-actors, tend to not demonstrate their emotions explicitly when they speak, thus making it harder to either recognize the emotion embedded in their spoken speech or distinguish one emotion from another by analyzing their speech. Appendix A captures brief emotion ontology from the Machine Learning perspective.

In spite of all the challenges enumerated in the previous chapter, researchers have been working to overcome these challenges to recognize emotions in spontaneous speech. The first comparative emotion recognition challenge was held in 2009 [99]. The interest in this area is mostly because of the importance in terms of its commercial applicability which can be seen in the number of patents filed in this area. Spontaneous speech emotion recognition finds use in several areas as shown in Fig. 2.1. It has been almost one and a half decade since people started to explore this area. Most recently in [78], authors have presented a method to classify fixed-duration speech windows as expressing anger or not (two-class problem). The proposed method does not depend on converting speech to text using a speech recognition engine. They also introduced the task of ranking a set of spoken dialogues by decreasing percentage of anger duration, as a step toward helping call center supervisors and analysts to identify conversations requiring further action.

Spontaneous speech emotion recognition is challenging as mentioned in the earlier chapter. The challenge is compounded due to noisy recording conditions and highly ambiguous ground truth labels (annotations). Further, noisy data infuses into the speech feature calculation process [65], thus making it difficult to find representative features which can be used to effectively train a classifier. Recently in [47], authors have tackled that problem by using deep belief networks (DBNs) that is well known for modeling complicated and nonlinear relationships among the acoustic

© Springer Nature Singapore Pte Ltd. 2017
R. Chakraborty et al., *Analyzing Emotion in Spontaneous Speech*,
https://doi.org/10.1007/978-981-10-7674-9_2

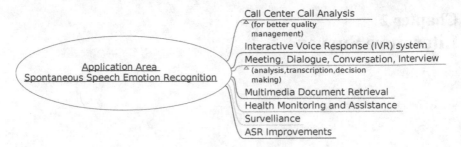

Fig. 2.1 Application of spontaneous speech emotion recognition

features. They evaluated HMM-DBN-based hybrid classifiers on FAU AIBO database to achieve the baseline results.

In [10], authors presented a system to detect anger in day-to-day conversations.
To deal with the difficulties of creating a good training corpus and to avoid the degradation in performances of using cross corpus samples, use of a carefully combined
cross corpus data is proposed.

An emotion classification paradigm based on emotion profiles (EPs) has been
described in [63]. This paradigm interprets the affective information in spoken speech
by giving probabilities for multiple classes rather than giving it a single label. EPs
provide an assessment of the emotion content of an utterance in terms of a set of simple
categorical emotions like anger, happy, neutral and sad. The EP method is
claimed to have a performance of 68.2% on the IEMOCAP database [11]. The system
is also capable of identifying the underlying emotions among the highly confusion
classes.

A comprehensive review in [95] addresses different emotional databases, modeling of different emotions, emotion annotations in database creation, the unit of
analysis, and the prototypicality. They also discuss different audio features, classification techniques, robustness issues, evaluation methods, implementation criteria,
and system integration.

In [91], both linguistic and acoustic feature were exploited for anger classification. In terms of acoustic modeling, authors generated statistics from well-known
audio features, like pitch, loudness, and spectral characteristics. While ranking those
features, it was revealed that loudness of the audio signal and Mel-frequency cepstral
coefficients (MFCCs) seems most promising for all databases that were considered.
Bag-of-words and term frequencies have been used for language-related modeling.
Classification based on both acoustic and linguistic features were then fused at the
decision level to classify anger.

Spontaneous speech emotion recognition system often built using acted speech
samples fails when tested in real-life scenarios, mainly because of the mismatch train-
test conditions. In [106], authors presented results with popular classifiers (support
vector machines (SVMs), multilayer perceptron (MLP) networks, k-nearest neighbor
(k-NN), decision trees, Naive Bayes classifiers, and radial basis function (RBF)
networks) used so far in speech emotion recognition.

A setup and performance optimization techniques for anger classification using acoustic cues is described in [90]. The setup is evaluated on a variety of features and two databases of German and American English. Using information gain based ranking scheme, they also reduced the feature dimension from a large set of features. In their analysis, they state that both MFCC and loudness are the most promising audio features.

Algorithms that can process naturally occurring human affective behavior was developed in [117] from a psychological perspective using human emotion perception. Authors have fused audio, visual, and language-related information; where the visual fusion was based on expressions, body gestures, and head movements. They outline some of the scientific and engineering challenges that have to be overcome to build technology that can benefit human affect sensing. They also emphasize an important issue in terms of the availability of well-labeled training and test data.

Acoustic anger detection systems that cater to human–machine interactions can provide better usable interfaces which can provide better user experience. In [9], a comprehensive voice portal for data collection is described to collect realistic audio data. Their experiments have shown that in comparison with pitch and energy related audio features, duration measures do not play an important role.

An acoustic and lexical classifiers-based emotion recognition system is proposed in [109]. They have used a boosting algorithm to find the discrimination on arousal and valence dimensions. Acoustic classifiers were reported to perform significantly better than the lexical classifiers, especially on arousal dimension, and the opposite results were observed in valence dimension. And an improved classification results were obtained by fusing acoustic and lexical features.

With an aim to distinguish negative and non-negative emotions, in spoken language data obtained from a call center application, the authors [48] propose the use of acoustic, lexical, and discourse information for emotion recognition. Domain-specific emotions were detected using linguistic and discourse information.

Emotion recognition for five categorical emotions (fear, anger, neutral, sad and relief) from a real-world French Human–Human call center audio corpus has been discussed in [111]. Using a large number of acoustic features, in the absence of speech transcriptions, they obtained a detection rate of 45%. The use of orthographic transcription (disfluences, affect bursts, phonemic alignment) in addition to the best 25 features, they achieved an improvement of 11%, namely 56% of detection rate.

In [33], authors proposed a two-stream processing scheme for emotion detection. The framework is based on two recognizers where one recognizes emotion from acoustic features and the other from semantics. Two outputs are then fused to get the final recognized emotion, where the confidence from each stream is used to weigh the scores from two streams for the final decision. This technique is useful for spontaneous and realistic data (e.g., call center) that have semantics associated with the speech. The proposed method is evaluated on LDC corpus and on call center data.

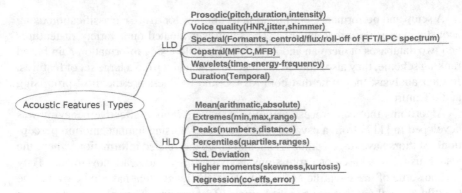

Fig. 2.2 Feature, both low-level and high-level descriptors, Commonly used in speech emotion recognition task

To correlate the commercial needs with the spontaneous speech emotion detection, authors in [98] discussed the main aspects of realistic and acted speech, effect of noises, channels, and speakers. They took three different databases into consideration, Berlin emotional speech database, Danish emotional speech database, and AIBO emotion corpus.

In an interesting study reported in [7], authors experimented with both elicited speech and speech acquired through WoZ scenario, where users believe that they are in conversation with dialogue system. It was observed that for the same feature set, namely the prosodic features, the closer the people got to a realistic scenario, the less reliable was the prosody feature an indicator of the emotional state of the speaker. They suggest the use of linguistic behavior such as the use of repetitions in addition to prosodic features for realistic audio utterances.

Emotion recognition from short utterances typically for interactive voice response (IVR) applications is explored in [116]. They aim to distinguish between anger and neutral speech, which is important in a commercial call center setup. They use a small database recorded by 8 actors expressing 15 emotions in their experiments. Results were compared for neural network, SVM, KNN, and decision trees.

An emotion recognition algorithm based on artificial neural network has been proposed in [64]. More importantly, they also proposed a method to collect a large speech database that contains emotions. They obtained recognition accuracy (around 50%) for 8 emotions for speaker-independent tests with their collected dataset. They suggest that emotion recognition to be a key component in a man–machine interaction.

Experiments on spontaneous speech emotion recognition has been carried out in [86]. Experiments were conducted on 700 short utterances of five emotions, namely anger, happy, neutral, fear, and sad. They used features like pitch, first and second formants, energy and speaking rate. They use feature selection technique for feature reduction, and then trained a neural network using the back propagation algorithm.

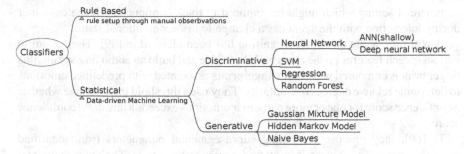

Fig. 2.3 Different classifiers used in speech emotion recognition

Like in automatic speech recognition, feature extraction is an important step in automatic speech emotion recognition. The performance of any pattern recognition system heavily depends on the features extracted from the raw signals; meaning how representative the extracted features are with respect to the raw signal and how discriminative they are in the higher dimensional space so that the classifier is able to use these characteristics to learn to discriminate emotions. In literature, researchers have experimented with different audio features (see Fig. 2.2) to gain a handle on spontaneous speech emotion recognition. Similarly, researchers have used different classifiers (see Fig. 2.3) for spontaneous speech emotion recognition. The classifiers can be broadly divided into two categories, namely

- Rule based | Rules are setup manually through observations
- Statistical | Machine learning data driven

  – Discriminative (SVM, ANN, Regression)
  – Generative (GMM, HMM, Naive Bayes)

2.2 Brief Survey of Patents in Audio Emotion Recognition

There is a rich patent-related literature in the area of audio emotion recognition. The richness, in the intellectual property space can be attributed to the commercial value associated with the problem of audio emotion recognition. In the following section, we briefly review the patents in this area in a chronological order. Note that there is no significant technical aspect that is mentioned in most patents, they usually describe a system or a method in a context to determine or recognize emotion from audio utterances.

Emotional state change in an audio signal is addressed in [23]. They first segment the audio into sequential segments and analyze each segment and determine the emotion state of the segment along with a confidence score. They further analyze the sequence of audio segments to identify if there has been a change in the identified emotion from previously identified emotion. This processing is important in a commercial setting which might be required to track continuously if the customer during interaction with the agent has a change in his/her emotional state.

A real-time audio emotion recognition has been claimed in [39]. They claim to extract speech features on the users mobile device and build an audio fingerprint, this finger print is compared with audio fingerprints associated with predefined emotions to determine relative degrees of similarity. They use a threshold to determine whether a confidence score for one or more particular emotions exceeds a threshold confidence score.

In [69], they propose computing suprasegmental parameters from identified acoustic phones and pauses in a uttered sentence by a user. The suprasegmental parameters is then used to calculate a language proficiency rating and the emotional state of for the use.

A method for real-time emotion detection in audio interactions between a customer and an agent is proposed in [46]. They extract feature vectors from speech utterances which are used to compare in the statistical space with modeled emotion labels. The emotion score represents the probability that the speaker that produced the speech signal is in certain emotional state.

In [20], the authors propose a mechanism to analyze the call center conversation between an agent and a customer to assess the behavior based on the voice of the customer. They first separate the customer and the agent voice and then apply a linguistic-based psychological behavioral model to determine the behavior of the agent and the customer separately.

In [8], a prosody analyzer is used to interpret natural language utterances. They propose to train the prosody analyzer with real-world expected responses which they claim improves emotion modeling and the real-time identification of prosody features such as emphasis, intent, attitude, and semantic meaning in the speaker's utterances.

The patent [38] talks of an emotion recognition system whose performance is independent of the speaker, the language or the prosodic information. The system is based on recognizing the phonemes and the tone of each phoneme from the speech signal. This information (phoneme + tone) is used to determine the emotion in the spoken utterance.

Emotions from the features extracted from voice using statistics is described in [85]. They assume the existence of a database which has associations of voice parameters with emotions. An emotion is selected from the database based on the comparison of the extracted voice feature to the voice parameters of all emotions.

Using the well-known statistical and MFCC features [44] proposes to recognize emotion from speech signal. This is achieved by scoring the speech sample with a known reference speech samples having different emotions. A probable emotional state is assigned to the speech signal based on the score.

In [40], a speaker's neutral emotion is extracted using a Gaussian mixture model (GMM) and all other than neutral emotions are classified using the GMM to which a discriminative weight for minimizing the loss function of a classification error for the feature vector for emotion recognition is applied.

The emotional state of a speaker at a given instant of time is discussed in [107]. They identify the affective states, namely arousal and valence from an utterance and associate it to an emotion label.

In [84], the authors suggest the use of fundamental frequency, standard deviation, range, and mean of the fundamental frequency in addition to a variety of other statistics as speech features to train a classifier that is able to assign an emotional state from a finite number of possible emotional states along with a confidence score.

Detecting an emotional state of a speaker is proposed in [81] based on the distance in voice feature space between a person being in an emotional state and the same person being in a neutral state. Their method is speaker-independent as no prior voice sample or information about the speaker is required.

Using intensity, tempo, and intonation in each word of a voice, [61] provides a technique to detect emotion of the speaker. They use the information about the speaker to identify the emotion of the person. This is a person dependent system and is able to identify emotions like pleasure, anger, and sadness.

The system proposed in [72] low-pass filters the voice signal before extracting voice features. These features, mostly statistical features based on the intensity and the pitch of the signal are then used to identify the emotion.

A system is proposed in [41] to monitor the emotional content of voice signals due to channel bandwidth. The system allows a person to observe the emotional state of the person on the other end of the telephone in a visual format.

A method for detecting an emotional state of a speaker based on his speech is mentioned in [80]. The approach is based on the computed distance between a person being in an emotional state and the same person being in a neutral state and similar to [81]. The neutral emotional state is learnt during training phase. However, as it happens in many cases the training and the testing environments are different leading to poor accuracy in the recognized emotional state.

In [55], a method is suggested that is based on determining phonemes in spoken utterances which is used for computing the voice quality features which are then used to improvement the emotion recognition system. Some of the features suggested are average normalized formant differences between the first two formants.

Emotional status of a speaker is determined in [49] using the spoken utterance of the speaker by using the intonation information. Essentially, they use the duration of flatness of the intonation in the speech signal to compute the emotion.

Patent [54] uses linguistic correlation information (LCI) to detect the emotional state (ES) of a given speaker based on their spoken speech. The LCI is derived by recognizing both the semantic and syntactic cue from the spoken utterance.

In [83] the authors propose a mechanism for monitoring a conversation between a pair of speakers (agent and customer) to determine the emotion of the customer. They first extract features from the voice signal, which includes the maximum value, standard deviation, range, slope, and mean of the fundamental frequency; also the mean of the bandwidth of the first and second formant; standard deviation of energy, speaking rate. These features are used to detect the emotion of the customer which they feedback to the agent to enable the agent to act appropriately.

A method of backward and forward inverse filtering on the uttered speech is used to detect the emotion of the speaker in [59]. They specifically obtain an estimate of the glottal wave which is used to ascertain the emotion content of the spoken utterance. The glottal wave is estimated by first performing backward and forward inverse filtering to obtain residual waveforms and then computing the cross correlation followed by integrating with maximum overlapping the backward and forward residual signal.

2.3   Brief Survey of Emotion Databases

Survey of available emotion databases useful for machines to learn emotions, suggests three broad categories [3, 9, 12, 79], namely (1) acted, (2) induced, and (3) spontaneous. Figure 2.4 shows a snapshot of a few well known, often used and easily accessible speech and audiovisual databases (chronologically in time) [110].

Generally for creating acted databases professional actors are asked to read and record spoken utterances assuming them to be in a certain emotional state. As mentioned earlier, speech emotion recognition systems perform well when acted data is used for creating training models [37, 92] because the speech corresponding to different emotions is very distinct and additionally these datasets have been recorded in a studio like environment. Most of the initial research was conducted on acted speech which produced good emotion recognition accuracies. This success can be attributed to the acted speech being more expressive, exaggerated, and recorded by professional and trained actors.

With the growing popularity of machine learning (ML) algorithms, they are being used for audio emotion recognition. Typically, ML algorithms create models through a training process using annotated data samples. It is essential to create emotion databases that can aid machines to analyze and learn human emotions effectively form audio corpus. There has been a considerable amount of work done in creating multimodal corpus consisting of emotionally rich human behavior [24, 26]. The latest and the most extensive corpus for acted speech can be found in Geneva multimodal emotion portrayals (GEMEPs) and audiovisual depressive language corpus (AViD-corpus).

GEMEP corpus consists of more than 7000 audio-video emotion portrayals, representing 18 emotions, portrayed by 10 professional actors who were coached by a professional director. These portrayals were recorded with optimal digital quality in multiple modalities, using both pseudolinguistic utterances and affect bursts. In addition, the corpus includes stimuli with systematically varied intensity levels, as well as instances of the masked expressions. From the total corpus, 1206 portrayals were selected and submitted to a first rating procedure in different modalities to establish validity in terms of inter-judge reliability and recognition accuracy. The results showed that portrayals reached very satisfactory levels of inter-rater reliability for category judgment [5].

AViD-corpus contains 340 video clips of 292 subjects performing a human–computer interaction task while being recorded by a webcam and a microphone. Each subject is recorded between one and four times, with a period of two weeks between each recording. The distribution of the recordings were 5 subjects appear in 4 recordings, 93 in 3, 66 in 2, and 128 in only one session. The length of these clips ranges from 20 to 50 min (average 25). The total duration of all clips is 240 hours.

Among the other prominent databases in acted category, Berlin database for emotional speech (Emo-DB) represents, unimodal speech only, acted in German and focused on 6 emotions that are considered as primary emotions. Ten actors of which 5 were male and 5 female were asked to read the same set of sentences in each of the five different emotions (anger, boredom, disgust, happy, sad, neutral, anxiety). Databases of this kind formed the backbone of early work on emotion. This database is freely available for usage (see [27]).

Danish emotional speech database (DES) was a part of VAESS project and had been recorded at Aalborg University. DES database consists of recordings from 2 male and 2 female actors, from radio theaters, expressing the 5 emotions, namely anger, happy, sad, surprise, and neutral. Each utterance duration is of 30 sec. The recordings use same utterances, expressed in different emotional situations to have emotional speech with less deviations from natural speech. The database is made up of 2 passages of fluent speech, 9 sentences, and 2 single words. This database is also freely available (see [28]).

Groningen (ELRA S0020) contains over 20 h of speech data collected from 238 speakers. It is a corpus of read speech material in Dutch. It has 2 short text and 1 long text. Orthographic transcription of the spoken utterances is also included. This database is partially oriented to emotion and is commercially available [32].

Interactive emotional dyadic motion capture (IEMOCAP) contains approximately 12 h of English audiovisual data recorded from 10 actors of which 5 are male and 5 female, including video, speech, motion capture of face, text transcriptions. It consists of dyadic sessions where actors perform improvisations or scripted scenarios, specifically selected to elicit emotional expressions. This dataset caters to 4 emotions anger, happy, sad and neutral, and can be freely downloaded [11].

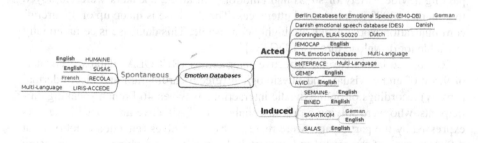

Fig. 2.4 Existing emotion databases

Ryerson multimedia lab (RML) emotional database contains 720 audio-visual emotional samples. This database caters to 6 basic emotions like, happy, sad, anger, fear, surprise and disgust. The recordings were done using digital video camera using a simple bright background and in a quiet environment. A list of emotional sentences were shared with the subjects and were directed to act and express their emotions naturally. A total of 10 sentences from each emotional class were provided. RML database is independent of language and cultural background. Emotional video samples from 8 subjects were collected. These subjects were selected on the basis of 6 languages they spoke, namely Mandarin, Punjabi, Persian, Urdu, English, Italian. The duration of each video clip was about 3–6 s with single emotion. This database is freely available through a license agreement [115].

eNTERFACE is a synthesized emotional speech database, recorded in a peaceful environment containing 150 to 190 utterances for each of the 6 emotions (anger, sad, joy, surprise, disgust and fear) proposed for video analysis in 4 languages (Slovenian, Spanish, French and English). 1 male and 1 female professional actors have recorded utterances for each language. This database is also available for commercial purpose [56].

With the increasing number of call centers, the interest of the research communities had shifted to spoken emotion recognition in spontaneous speech [15, 16]. But performance of the same spoken emotion recognition system, trained using acted data, drastically degrades when tested on spontaneous utterances, since emotions expressed in realistic spontaneous speech are often subtle. However, researchers tackled the difficulties in recognizing emotions in spontaneous speech by using data-driven approaches, i.e., training the classifier using annotated data [43, 74]. Creating such annotated databases itself is a very challenging task since it involves lots of human interventions, efforts, time and cost.

Among the prominent and publicly available spontaneous emotional speech databases, HUMAINE is an audio-video database of 50 video clips and is made up of natural as well as induced emotional data. The duration of the recorded data ranges from 5 s to 3 min. This database is available from [25].

Another popular spontaneous speech database, called SUSAS, was collected for analysis of speech in noise and stress. The database consists of five domains, encompassing a wide variety of stress and emotions. In all 32 speakers were employed to generate 16000 isolated word utterances. The database is made up of 35 aircraft communication words which are highly confusable. This database is commercially available for download through LDC [36].

Remote collaborative and affective interactions (RECOLA) database consists of 9.5 h of audio, visual, and physiological (electrocardiogram and electrodermal activity) recordings of online dyadic interactions between 46 French-speaking participants, who were solving a task in collaboration. Affective and social behaviors expressed by the participants were reported by themselves (encoded emotions), at different steps of the study, and separately by 6 French-speaking assistants using a Web-based annotation tool. The dataset was split into three partitions (training, development, and test). This database is available on requested through an end user license agreement [93].

LIRIS-ACCEDE is an annotated emotional database created under Creative Common (CC) License, and contains 9800 video clips, segmented from 160 short films, having total duration of 73 hrs and 41 min. 8–12 s duration video clips were used to conduct emotion recognition experiments. A robust shot and fade in/out detection was used, to make sure that each extracted video clip start and end with a shot or a fade so that temporal transitions of emotions can be studied. This database is freely available by signing a license agreement [102].

Real-world spontaneous spoken utterances can be obtained during a human–human conversations (e.g., call center call conversations, meetings, chats, talk shows, interviews.) for creating a realistic spontaneous speech databases [16, 22, 74]. But in scenarios, especially the call center call conversations, conversation carry personal confidential information which imposes legal constraints on its creation and use. For this reason, there are very few spontaneous databases which are publicly available. This has motivated researchers to simulate an environment to evoke emotions from the subjects which are considered to be semi-spontaneous in nature (also popularly known as induced in the literature) [22, 43].

In induced databases, subjects are asked to read emotive texts, which are expected to naturally induce the actual emotion in the speaker [101, 102]. SEMAINE corpus is one such database which consists of emotionally colored conversations. Subjects were recorded when in conversation with an operator who plays four different roles to evoke specific emotional reaction from the subject. The operator and the user were seated in separate rooms, where they could see each other through teleprompter screens, and hear via speakers. This interaction is captured by 5 high-resolution, high frame-rate cameras in addition to 4 microphones. Continuous annotations in 21 dimensions including `valence` and `arousal`, by three to seven raters per session is included along with transcription is included as part of the database. This database can be obtained from [58].

The Belfast natural-induced emotion database (BINED) contains samples of mild to moderately strong naturalistic emotional responses, which were based on a series of laboratory tasks. Each recording is a short video clip with stereo sound. The task were based on fixed context designed to elicit an emotional state. The database includes annotations for both encoded and decoded emotions. This database can be download by signing an agreement [104].

SMARTKOM—Bavarian archive for speech signals multimodal corpora is a collection of telephonic conversational data where the conversation is led between the wizard and operator in what is called the wizard operator test (WOT) mode. The database consists of 22 recording sessions collected from 43 participants (25 males and 18 females), with an average duration of 4.57 min. This database is available for public usage at [6].

In this section, we have only summarized acted, spontaneous, and induced emotion databases that are available and widely used by research community (see Appendix C). However, there are many more emotion specific databases that have been used by researchers in their work (see [24, 105]). A comprehensive list of databases suitable for audio emotion recognition is available at [19].

2.4   Challenges in Building an Audio Emotion Database

One of the main difficulty in creating an audio database for emotion research stems from the fact that you need a person to be in a certain emotional state and then you need him to speak or utter something in that state. Not all emotional states elicit speech, for example when one is `sad`, it is very rare for someone to speak. Creating spontaneous emotion databases have challenges in terms of some emotions not being captured in the database and some emotion only being captured in the presence of noisy environment.

Skewed. Spontaneous speech data which can be labeled `sad` is very difficult to create in a natural setting because it is something that is not elicitable speech from people. As a result, capturing `sad` emotional audio data from spontaneous speech is not feasible, making the database very skewed in favor of only those emotions where the person tends to speaks, like `anger`.

Noisy. Certain emotions, like `happy`, are expressed in the form of spoken speech more often in a social gathering, in which case there is a lot of babble noise.

For this reasons, to build a good audio database suitable for emotion analysis is very challenging. Note that to gain access to spoken audio, it is required for a person to speak. In general, a person speaks when there is a listener (another human or a machine) or an audience. So to generate a situation which allows a person to speak (audio) there is a need for at least one *more* person or an intelligent machine for the person to have a dialogue with.

Voice-based call centers (where there is a natural dialogue between a customer and an agent) are a perfect setting to collect data. There are certain challenges in capturing audio conversation between an agent and a customer in a call center.

- Agent spoken data is emotionally constrained because they are trained to not express certain emotions (like `anger`).
- It is observed that customers rarely speak to express happiness with a service, they only call when there is a deficiency in the service, which in turn means the customer is emotionally either `neutral` or `angry` and most likely not when `happy`.
- Most of the conversations have customer specific privacy data, which brings in legal issues in terms of recording speech and distributing it.

The best approach to collect speech data with emotion seems to be to engage a speaker in an intelligent dialogue which allows the speaker to experience a certain emotion and simultaneously allow him to speak. For example,

- An *intelligent* system is aware that the ambient temperature is warm and during a dialogue with a human poses a question `"Isn't it very cold today?"` which would most likely lead to an utterance or an expression which captures `surprise`. This can be recorded.

- Music is known to induce a certain kind of emotion. Play a music to induce a certain emotion and then elicit the person to speak something which can be recorded.

Considering the embedded challenges for creating suitable emotional speech database, we have made an attempt to collect audiovisual emotional data that is close to being spontaneous.

Chapter 3
A Framework for Spontaneous Speech Emotion Recognition

3.1 Overview

Automatic spontaneous speech emotion recognition is an important aspect of any naturalistic human–computer interactive system. However, as emphasized in earlier chapters, emotion identification in spontaneous speech is difficult because most often the emotion expressed by the speaker is not necessarily as prominent as in acted speech. In this chapter, we elaborate on a framework [18] that is suitable for spontaneous speech emotion recognition. The essence of the framework is that it allows for the use of contextual knowledge associated with the audio utterance to make it usable for emotion recognition in spontaneous speech. The framework is partly motivated by the observation [15] that there is significant reduction in disagreement among human annotators when they annotate spontaneous speech in the presence of additional contextual knowledge related to the spontaneous speech. The proposed framework makes use of these contexts (for example, linguistic contents of the spontaneous speech, the duration of the spontaneous speech) to reliably recognize the current emotion of the speaker in spontaneous audio conversations.

3.2 The Framework

In general, the problem of identifying emotion for a speech signal $s(t)$ of duration T seconds is posed as that of assigning a label $\varepsilon \in \mathscr{E}$ to the speech signal $s(t)$, where

$$\mathscr{E} = (E_1 = \text{anger}, E_2 = \text{happy}, \cdots, E_n)$$

is the set of n emotion labels. Typically, the emotion label of a speech signal $s(t)$ is assigned by first computing the posterior probability or score associated with $s(t)$ being labeled as emotion E_k $\forall k = 1, 2, \ldots, n$ as

$$\epsilon^p_{k,s(t)} = P(E_k|s(t)) = \frac{P(s(t)|E_k)P(E_k)}{P(s(t))} \tag{3.1}$$

© Springer Nature Singapore Pte Ltd. 2017
R. Chakraborty et al., *Analyzing Emotion in Spontaneous Speech*,
https://doi.org/10.1007/978-981-10-7674-9_3

where $\epsilon^p_{k,s(t)} = P(E_k|s(t))$ is the posterior probability of $s(t)$ being labeled as emotion E_k.

Note that (a) the posteriors are learnt from a reliably labeled training speech corpus using some machine learning algorithms and (b) in practice, the features extracted from the speech signal $f(s(t))$ are used instead of the actual raw speech signal $s(t)$ for training a machine learning system. The identified emotion $E_{k*} \in \mathscr{E}$ of the speech signal $s(t)$ is the posterior with maximum score, namely

$$E_{k*} = \arg \max_{1 \leq k \leq n} \{\epsilon^p_{k,s(t)}\} \tag{3.2}$$

This conventional process of assigning an emotion label (3.2) to a speech signal $s(t)$ works well for acted speech, because (a) the entire speech utterance, $s(t)$ of duration T carries one emotion and (b) there exists a well-labeled speech corpus available. However, both these conditions, namely the entire speech signal carrying a single emotion and the existence of a reliably labeled corpus, are seldom true for spontaneous or conversational speech. For example, in a call center conversation between the agent and a customer, the emotion of the customer can change significantly during the length of the conversation. As mentioned earlier, the emotions in spontaneous speech are not explicitly demonstrated and hence cannot be robustly identified even by human annotators (especially in the absence of contextual information surrounding the speech utterance) in order to build a emotion labeled speech corpus.

These differences between the acted and the spontaneous speech (highlighted in detail in Chap. 1) motivates us to identify a suitable framework that can work for recognizing emotions in spontaneous speech [15] in addition to being usable for identifying emotion of acted speech. The framework tries to handle scenarios like (a) the emotion of the entire speech signal need not be the same and (b) relies on the fact that human annotators are better able to recognize emotions when they are provided with the associated contexts of the speech signal.

The working of the framework is driven on the premise that emotions do not vary within a small duration speech signal and that it is indeed possible to compute reliably the emotion of a smaller duration speech signals [73]. Let $s_\tau(t)$ be a segment of $s(t)$ at $t = \tau$ and of duration 2Δ, where $\Delta \leq T$. Namely,

$$s_\tau(t) = s(t) \times \{U(\tau - \Delta) - U(\tau + \Delta)\}$$

where $U(t)$ is a unit step function defined as

$$U(t) = 1 \quad \text{for} \quad t \geq 0$$
$$= 0 \quad \text{for} \quad t < 0.$$

Now instead of computing the emotion for the complete signal, namely $s(t)$, the framework computes the emotion of smaller speech segments, namely $s_\tau(t)$. As can be observed, (a) $s_\tau(t) \subset s(t)$ and is of length 2Δ and (b) $\tau \in [0, T]$. Also, note that the number of speech segments is dependent on T and Δ and can be easily computed as $\lfloor \frac{T-\Delta}{2\Delta+1} \rfloor$ under the assumption that the speech segments do not overlap. In the framework for spontaneous speech, we not only compute the emotion as mentioned earlier, albiet for the short speech segment $s_\tau(t)$, namely

$$\epsilon^p_{k,s_\tau(t)} = P(E_k|s_\tau(t)) \tag{3.3}$$

for $k = 1, 2, \ldots n$ emotion labels, but also make use of the emotions computed from the previous η speech segments, namely $\epsilon^p_{k,s_{\tau-v}(t)}$ for $v = 1, 2, \ldots \eta$. So we have the posterior score associated with the speech utterance $s_\tau(t)$ being labeled E_k as

$$'\epsilon^p_{k,s_\tau(t)} = \epsilon^p_{k,s_\tau(t)} + \sum_{v=1}^{\eta} \omega_v \epsilon^p_{k,s_{\tau-v}(t)} \tag{3.4}$$

where $\omega_1, \omega_2 \cdots, \omega_\eta$ are such that $\omega_i < 1$ for $i = 1, 2, \cdots, \eta$ and $\omega_1 > \omega_2 > \cdots > \omega_\eta$. Note that the posterior score of the speech segment $s_\tau(t)$ is influenced by the weighted sum of the posterior score of the previous speech segments as seen in (3.4). This is generally true of spontaneous conversational speech where the emotion of the speaker is based on the past emotions experienced by the speaker during the same conversation.

Further, the output posterior scores from emotion recognizer, namely $\epsilon^p_{k,s_\tau(t)}$ is modified by a knowledge module; the posterior scores are appropriately weighted based on the magnitude of τ which is an indication of the temporal distance of the speech segment from the start of the call. This can be represented as,

$$\epsilon^\kappa_{k,s_\tau(t)} = w_\tau \epsilon^p_{k,s_\tau(t)} \tag{3.5}$$

where $\epsilon^p_{k,s_\tau(t)}$ and w_τ (see Fig. 3.1) are the posterior probability score and a weight vector at time instant τ, respectively. And $\epsilon^\kappa_{k,s_\tau(t)}$ is the emotion computed based on the knowledge of the time lapse between the segment $s_\tau(t)$ and the start of the call $s(t)$. The use of temporal position of the speech segment is motivated based on the observation that the length of the customer-agent conversation plays a very important role in influencing a possible change in the customer's emotion.

As mentioned in [15], w_τ is expected to grow or decay exponentially as a function of τ. The growth and decay depend on the type of the emotion. As an example (see Fig. 3.1), it is expected that w_τ for emotions like anger and sad closer to the end of the conversation is likely to be more intense compared to the same emotion of the customer at the beginning of the call. Subsequently, as seen in Fig. 3.1, the weight w_τ is expected to grow as a function of (τ) time for anger and sad while at the same time w_τ is expected to decay as time index increases for happy emotion.

Our framework combines ϵ^p_k and ϵ^κ_k to *better* estimate the emotion of the spontaneous utterance $s_\tau(t)$ as

$$e^\kappa_k = \lambda_p \cdot ('\epsilon^p_k) + \lambda_\kappa \cdot (\epsilon^\kappa_k) \tag{3.6}$$

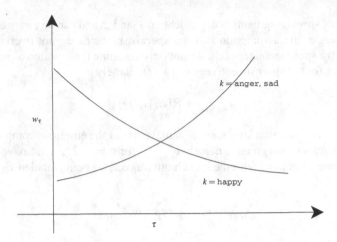

Fig. 3.1 Knowledge regarding the time lapse of the utterances in a call. The weights w_τ of emotions like happy decreases with τ while the weights increase for emotions like anger, sad

where $\lambda_\kappa = 1 - \lambda_p$ and $0 \leq \lambda_\kappa, \lambda_p \leq 1$. Clearly when $\lambda_\kappa \neq 0$, the framework makes use of the contextual knowledge and in this scenario, the emotion of the spontaneous speech utterance $s_\tau(t)$ is computed as

$$E_{k*} = \arg \max_{1 \leq k \leq n} \left\{ e_k^\kappa \right\} \tag{3.7}$$

The proposed framework for spontaneous speech recognition which uses contextual information is shown in Fig. 3.2. Knowledge regarding the time lapse of the utterance in an audio call, especially in conversational transactions like call centers, provides useful information to recognize the emotion of the speaker. Subsequently, our framework incorporates this knowledge in extracting the actual emotion of a user.

It is often observed that speaker, especially the agents in a call center, is often trained to suppress their emotions during their conversational with their customers inspite of using emotion laced linguistic words in their spoken sentences [17, 48]. This is also observed sometimes in spontaneous speech. We expand our framework by incorporating this observation into our framework. Incorporating linguistic contextual knowledge is an important part of the framework used to recognize emotion in spontaneous speech. This is useful especially in situations where it might be difficult to compute emotions solely based on the acoustics of the speech signal or distinguish between two emotions because they are close in terms of their posterior scores.

To identify emotionally salient words spoken in an utterances, an information-theoretic concept of emotional salience has been adopted in [48]. Generally, the linguistic words are mapped to a emotion label based on how frequently they occur in context of that emotion label. For example, the word "nice" is more likely to occur in a happy emotion but it can also sometimes occur in an angry emotional state (e.g.,

Fig. 3.2 Associated with time lapse

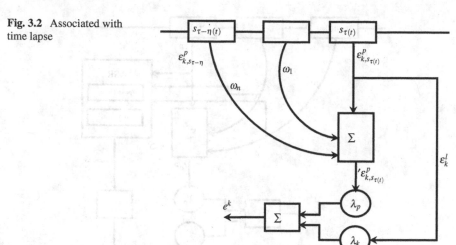

"this is not a nice way to treat your clients", (see Fig. 1.5)). Emotional prominence associated with spoken utterances is represented as ϵ_k^l and used to spot automatically the emotionally salient words. Salience is a probabilistic measure that is used to find and associate words that are related to emotions in the speech data. We combine ϵ_k^p and ϵ_k^l into our framework which in turn enables the framework to address spontaneous and conversational speech in terms of being able to better estimate the emotion of spoken utterance $s_\tau(t)$, namely

$$e_k^l = \lambda_p \cdot' \epsilon_k^p + \lambda_l \cdot \epsilon_k^l \tag{3.8}$$

where $\lambda_l = 1 - \lambda_p$ and $0 \leq \lambda_l, \lambda_p \leq 1$. The framework makes use of contextual linguistic knowledge when $\lambda_l \neq 0$. Emotion of the spontaneous speech utterance $s(t)$ with the incorporation of linguistic knowledge (ϵ_k^l) is represented as

$$E_{k*}^l = \arg \max_{1 \leq k \leq n} \{e_k^l\}. \tag{3.9}$$

The complete framework for spontaneous speech emotion recognition is shown in Fig. 3.3. The main advantage of our framework is its scalability in terms of its ability to incorporate a number of contextual knowledge. One can add as many contextual knowledge sources that are available with the audio utterance without any limitation. In this way, our proposed framework can combine $\epsilon_k^p, \epsilon_k^\kappa$, and ϵ_k^l to give better output as shown in Fig. 3.4. We can generalize the framework as

$$e_k^f = \lambda_p \cdot' \epsilon_k^p + \lambda_\kappa \cdot \epsilon_k^\kappa + \lambda_l \cdot \epsilon_k^l \tag{3.10}$$

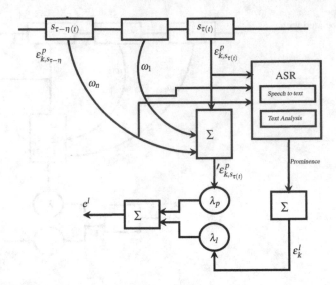

Fig. 3.3 Proposed framework incorporating linguistic knowledge

Fig. 3.4 The complete
framework for spontaneous
speech emotion recognition

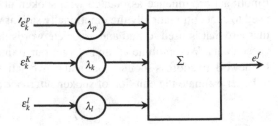

and the identified emotion being

$$E_{k*}^f = \arg \max_{1 \le k \le n} \left\{ e_k^f \right\}. \tag{3.11}$$

3.3 Framework Validation

Experiments were conducted on both spontaneous speech datasets [1, 2] and acted
datasets (Emo-DB) [27] to check the validity and appropriateness of the proposed
framework. Detailed description of the spontaneous dataset is provided in Table 3.1.
Using Fleiss' Kappa statistics (which is used for multiple labelers) [57, 112], we
obtain a kappa score β in the range of 0.68 and 0.70, which corresponds to substantial inter-labeler agreement when the annotators were provided with the additional information about the audio calls. In all our validation experiments, we used

Table 3.1 Details of spontaneous speech dataset

Domain	IVR- SERES	CALL CENTER
Total no. of calls	117	210
Total duration (min)	308	850
Avg no. of user utterances (turn) per call	18	27
No. of annotators	7	7
Kappa score without knowledge	0.14	0.17
Kappa score with knowledge	0.68	0.70
Data used for experiments		
Total no. of utterances	1265	2118
Avg utterances length (s)	3.1	11
No. of utterances (Training; 80%)	1012	1694
No. of utterances (Testing; 20%)	253	424

utterances having $\beta \geq 0.6$. The number of utterances is 1264 and 2117, respectively, for IVR- SERES and CALL CENTER respectively.

Training–testing sample distribution was maintained in the 80-20% ratio for all experiments, and data balancing was done so that the classifiers are trained uniformly for all emotion classes. All the audio samples in our experiments are sampled at 8 kHz, 16 bit, and monoaural. In our initial study in [15], we considered the complete duration of the utterances for extracting emotions; however, in our later experiments, we used $2\Delta = 400$ msec [16, 17, 73].

OpenSMILE paralinguistic 384 dimension audio features [99] are extracted and reduced to a lower dimension after feature selection using the WEKA Toolkit [108]. Different classifiers support vector machine SVM, artificial neural network (ANN), and k-NN have been used in our experiments. LibSVM toolkit is used for implementing SVM classifier [51]. For SVM, we used polynomial kernel and pairwise multiclass discrimination based on sequential minimal optimization. A three layered ANN was trained using Levenberg–Marquardt back propagation algorithm [34]. The number of neurons used in the hidden layer was selected through extensive experimentation. To be fair in our comparison, we report results for the four most frequently observed emotions in spontaneous call conversations [33, 82]) namely anger, happy, sad and neutral.

Table 3.2 shows the performance of the emotion recognition system on spontaneous (IVR- SERES and CALL CENTER) and acted (Emo-DB) speech datasets using three different classifiers and their combination. While testing the system on spontaneous datasets, we conducted two sets of experiments. In the first, emotion samples from the acted Emo-DB dataset were used to train the classifier, thus resulting in a train–test mismatched condition (represented as U in Table 3.2). We checked how the use of associated contextual knowledge improves the spontaneous speech emotion recognition performance even if the classifiers are trained with the samples from a different dataset (Emo-DB, acted speech corpus in our case). This situation may arise

Table 3.2 Emotion recognition accuracies (%) on spontaneous and acted datasets

IVR- SERES

Description	λ_p	λ_κ	λ_l	SVM		ANN		k-NN		SVM+ANN+k-NN	
				U	M	U	M	U	M	U	M
Conventional(baseline) ϵ_k^p	1	0	0	38.7	71.3	40.7	72.9.3	40.2	62.9	43.8	73.9
+time lapse (ϵ_k^κ)	1	1	0	47.2	73.1	51.2	76.5	39.8	66.1	55.4	76.1
+ linguistic contents (ϵ_k^l)	1	0	1	54.8	76.4	61.1	78.6	40.9	68.1	64.5	82.2
+time lapse + linguistic content	1	1	1	59.2	80.2	65	79.7	52.8	70.4	70.3	85.2

CALL CENTER

Description	λ_p	λ_κ	λ_l	SVM		ANN		k-NN		SVM+ANN+k-NN	
				U	M	U	M	U	M	U	M
Conventional(baseline), ϵ_k^p	1	0	0	38.1	69.8	39.2	68.4	34.2	64.2	42.9	72.6
+time lapse(ϵ_k^κ)	1	1	0	52.2	75.2	57.1	76.1	42.6	70.1	59.7	77.1
+ linguistic contents(ϵ_k^l)	1	0	1	53.1	75.8	52.1	75.9	40.5	67.7	38.4	80.2
+time lapse + linguistic content	1	1	1	58.1	80	54.5	76.6	48.1	71.7	68.2	82.1

Emo-DB

Description	λ_p	λ_κ	λ_l	SVM	ANN	k-NN	SVM+ANN+k-NN
				M	M	M	M
Conventional(baseline) ϵ_k^p	1	0	0	85.8	84.1	79.8	89.1

if someone does not have the luxury of using annotated data because of the fact that annotation requires substantial amount of human intervention and cost and may not be available for call center calls because of the infrastructure issues. Conversely, in the second set of experiment, emotion samples from the same dataset were used to train the classifier (M in Table 3.2 to represent *matched* condition). As can be seen, SVM classifier performs consistently best among all the classifiers. The classifier combination further improves the recognition accuracy.

To identify emotion by relying only on the output from emotion recognizer and not incorporating any knowledge, λ_p is set to 1, while λ_κ and λ_l are set to 0 (marked as conventional (baseline) in Table 3.2 for all the three datasets, namely IVR- SERES, CALL CENTER and Emo-DB). Note that, weight vectors (i.e., w_τ) are learnt separately from the available meta-data associated with the calls (transcriptions and associated time stamps of the spoken utterances). The vector is learnt from the training dataset by averaging the scores given by the human annotators for each emotion class. Emotional prominence is learnt from the text transcripts of the utterances. It is observed that the speakers tend to use some specific words for expressing theirs emotions. This was evident while listening to the audio used to annotate [88].

As can be seen in Table 3.2, the system that relies only on ϵ_p (i.e., when $\lambda_p = 1$) gives a low-performance accuracy for spontaneous speech dataset (best accuracy of 43.8% obtained by combining classifier for mismatched condition). This is expected, since the recognizer is trained with the acted speech samples, but tested with the spontaneous speech (a mismatch in training–testing). However, the accuracy improves by an absolute value of 30.1% in matched scenario. However, the results improve when the knowledge related to either time lapse of the utterance in the call ($\lambda_p, \lambda_\kappa \neq 0$) or the linguistic content ($\lambda_p, \lambda_L \neq 0$) is used. Finally, when both of these knowledge are used (namely $\lambda_p, \lambda_\kappa, \lambda_l \neq 0$), system produces the best recognition accuracies for mismatched condition; an absolute improvements of 26.5% and 25.3% are achieved for two spontaneous datasets (IVR- SERES and CALL CENTER), respectively. The best accuracies of 85.2% (for IVR- SERES) and 82.1% (for CALL CENTER) are achieved for matched scenario. Interestingly, for IVR- SERES dataset, more improvement was found when linguistic information is used in comparison to the time-lapse-related information. However, the improvement is small for the CALL CENTER calls when time-lapse-related information is used. We hypothesize that this could be because of the nature of the datasets; the utterances in IVR- SERES are more isolated compared to the utterances in a call center setup which leads to time-lapse-based knowledge contributing to lesser improvement in emotion recognition accuracies. We also observed that the performance accuracy on acted dataset is better compared to the spontaneous dataset. The classifiers were better learnt with acted speech, even with the small number of samples per class in comparison to the spontaneous speech. This might be due to the fact that acted samples were recorded in the controlled environment with minimum background noise, whereas the spontaneous speech utterances (especially from users) are from different backgrounds having variable types and levels of noise. To be consistent in our experimentations, we also tested the proposed framework using only either the time-lapse-related knowledge ($\lambda_p = \lambda_l = 0, \lambda_\kappa \neq 0$) or the linguistic knowledge ($\lambda_p = \lambda_\kappa = 0, \lambda_l \neq 0$), like we did in case of just relying on the

Table 3.3 Linguistic context improves emotion recognition in a IVR-SERES Call

Emotion Labels: A = anger, H = happy, N = neutral, S = sad

Sample Call (from IVR-SERES)	Scores using ε^p				Label	Scores using ε^l				Label	$\varepsilon^p + \varepsilon^l$
	$k=1$ (A)	$k=2$ (H)	$k=3$ (N)	$k=4$ (S)		$k=1$ (A)	$k=2$ (H)	$k=3$ (N)	$k=4$ (S)		
System: Welcome to the system											
System: Please select a language											
User: English	0.1	0.1	0.8	0.0	N	–	–	–	–	–	N (0.8)
System: Which functionality would you select?											
User: Seat availability	0.1	0.0	0.7	0.1	N	–	–	–	–	–	N (0.7)
System: Unable to understand	Event Occured										
User: Seat availability	0.2	0.5	0.2	0.1	H	1.0	0.0	0.0	0.0	A	A (1.2)
System: Which train?											
User: Train (x)	0.6	0.1	0.1	0.2	A	–	–	–	–	–	A (0.6)
System: Which station (from)?											
User: Station (A)	0.4	0.0	0.6	0.0	N	–	–	–	–	–	N (0.6)
System: Which station (to)?											
User: Station (B)	0.2	0.1	0.7	0.0	N	–	–	–	–	–	N (0.7)

emotion recognizer ($\lambda_\kappa = \lambda_l = 0$, $\lambda_p \neq 0$). For IVR- SERES calls, recognition accuracies of 32.3% and 43.7% are observed for the system that relies only on time-lapse related or linguistic knowledge, respectively. For call center calls, recognition accuracies of 35.2% is achieved by using only time-lapse-related knowledge and 40.7% with only the linguistic knowledge. From the experimental results, it can be said that the fusion of all available knowledge along with the emotion recognizer output improves the overall performance the system.

For speech to text conversion, instead of taking the ASR output, we considered the manual transcription for the spoken utterances to ensure that ASR errors do not propagate through the system. In Table 3.3, we demonstrate how the linguistic knowledge (i.e., e^l) corrects the output of the emotion recognizer. First column represents the manual transcription of the audio call conversation between an user and IVR computer (represented as "System" in Table 3.3). Second column represents the scores (for four classes) at the output of the emotion recognition system, and the fourth column represents the vectors at the output of linguistic knowledge-based system. The last column is the final output scores, which are obtained by combining the pairwise scores presented in $(2-5)^{th}$ and $(7-10)^{th}$ column. It is very hard to expect the emotion of an user to be happy when she has to repeat the same query (*seat availability*) for the second time due to the fact that system is not able to recognize what is spoken by the user (here, it is an event or context). Our framework allows for correcting the emotion by using the event (or context)-based knowledge (λ_p, $\lambda_l \neq 0$). The conventional system (namely, ϵ_k^p) outputs happy, whereas after combining the event-based knowledge (namely e^l), the output becomes anger.

3.4 Conclusions

Recognizing emotion in spontaneous speech is difficult because it does not carry sufficient intensity to distinguish one emotion from the other. In this chapter, we hypothesize that the lack of discriminating properties in the audio can be handled by making use of prior knowledge in the form of time lapse of the audio utterances in the call and linguistic contexts. The development of this knowledge-based framework for spontaneous speech emotion recognition is generic in the sense that it boils down to the conventional method of emotion recognition when $\lambda_\kappa = \lambda_l = 0$.

Experimental validations proved that the use of different context-dependent knowledge in terms of the time lapse of the utterances in the audio call and the linguistic content are important for emotion recognition in spontaneous speech. The framework has been evaluated on three different databases in both matched and unmatched train-test conditions. Incorporation of prior knowledge and fusion at the classifier level to better the recognition of emotion in spontaneous speech, even in an unmatched train-test scenario, clearly establishes the usefulness of the proposed framework [15].

Chapter 4
Improving Emotion Classification Accuracies

4.1 Overview

Identification of emotions in spontaneous speech has several challenges; some of them are already discussed in the previous chapters. We have already seen that the incorporation of contextual knowledge improves the performance of recognition system. Infact, the improvements can be achieved at different levels of recognition process, either at the frontend or at the backend. Figure 4.1 depicts the possibilities of the improvement at different levels of emotion recognition system. Specifically, those improvements can be done at three different levels, namely at signal level [76, 89], feature level [42, 100], and classifier level [13, 14]. In this chapter, we are going to discuss different methods to improve emotion recognition in spontaneous speech.

4.2 Signal Level: Segmentation of the Audio

There is a lot of debate on the minimum temporal length of the audio for which the emotions can be extracted reliably. The ideal length is mostly dependent on the application and largely varies with the contexts of the environment. While the approach adopted by most emotion recognition schemes described in the literature makes use of the complete audio, we propose a simple approach of segmenting the audio. Given an audio $s(t)$ of duration T, and an emotion recognition algorithm $\mathscr{E}$, and a set $E = e_1 = \texttt{anger}, e_2 = \texttt{sad}, \ldots, e_n$ consisting of n distinct emotion labels. The classical emotion recognition problem can be represented as

$$\mathscr{E}\left(s(t)\right) \longrightarrow E. \tag{4.1}$$

In literature, we find several methods proposed to represent $\mathscr{E}$ and very often instead of working directly on the signal $s(t)$, the idea is to extract some features (like pitch, mel-frequency cepstral coefficients) from the speech signal and use them instead. So

© Springer Nature Singapore Pte Ltd. 2017
R. Chakraborty et al., *Analyzing Emotion in Spontaneous Speech*,
https://doi.org/10.1007/978-981-10-7674-9_4

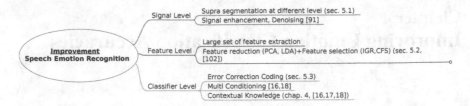

Fig. 4.1 Different possibilities for improved emotion recognition

(4.1) becomes

$$\mathscr{E}\left(\mathscr{F}\left(s(t)\right)\right) \longrightarrow E \qquad (4.2)$$

where $\mathscr{F}$ is some feature extracting process.

In our approach, we segment the audio $s(t)$ into smaller non-overlapping segments, namely we construct

$$s_k = s(k\tau, (k+1)\tau) \qquad (4.3)$$

for some $\tau > 0$ and $k = 0, 1, 2, \ldots, ((T/\tau) - 1)$. And compute $\mathscr{E}\left(\mathscr{F}\left(s_k\right)\right)$ for all k segments separately. Let

$$\varepsilon_k = \mathscr{E}\left(\mathscr{F}\right)(s_k)) \qquad (4.4)$$

where $\varepsilon_k \in E$ (namely ε_k can take a value e_1 or e_2 or $\cdots$, e_n) and is the emotion recognized for the speech segment s_k. We use all the ε_k's, namely $\{\varepsilon_k\}_{k=0}^{((T/\tau)-1)}$ to determine the emotion of the speech $s(t)$. The emotion associated with $s(t)$ is the emotion that occurs the maximum number of times in the set $\{\varepsilon_k\}_{k=0}^{((T/\tau)-1)}$. Our approach is shown in Fig. 4.2 along with the conventional approach.

In initial experiments, we use a rule-based emotion recognition system ($\mathscr{E}$) [94] to estimate the emotion from audio and the features ($\mathscr{F}$) that have been used are some basic speech features, namely pitch standard deviation (σ_p), pitch mean slope ($\bar{p}_s$), pitch average ($\bar{p}$), pitch first quantile (p_q), jitter (J), intensity standard deviation (σ_i) has been suggested in [94]. For all our experiments, we used $\tau = 400$ ms.

4.2.1 Experimental Validations

We conducted two sets of experiments. The first experiment was to identify the smallest segment of the speech that can reliably identify the emotion of a speech segment. This experiment was conducted to identify τ in (4.3). The second set of experiments were conducted to verify the performance of the proposed approach of segmenting the audio into smaller segments and fusing the identified emotions of each segment (see Fig. 4.2), and the third experiment was conducted on a real call center conversation, which is a natural spoken conversation between an agent and

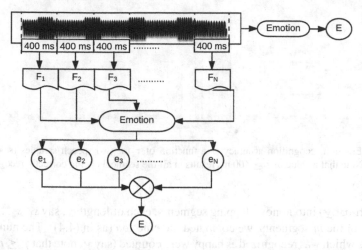

Fig. 4.2 Emotion estimation process using the complete audio (conventional approach) and segmented audio (proposed approach)

Table 4.1 Emotion recognition accuracy (%) conventional approach

Emotions	Conventional approach (Fig. 4.2)	
anger	(82/127)	64.50
happy	(38/71)	53.50
neutral	(34/79)	42.00
sad	(15/66)	22.75
Total	(169/343)	49.20

a customer with varying emotion during the call. We used Emo-DB database [27] for the first two experiments while we used a real call center conversation from a financial institute for our third experiment.

For all our experiments, we considered only four emotions, namely anger, happy, neutral and sad because the rule-based emotion extraction system ($\mathscr{E}$) [94] that we used catered to only these four emotions. There were 343 audio files corresponding to these 4 emotions, specifically, 127 emoting anger, 71 emoting happy, 66 emoting sad and 79 emoting neutral. All experimental results are reported on this data set. Table 4.1 shows the performance of the rule-based emotion recognition system on the 343 audio files. Clearly, the recognition accuracy of the emotion recognition system $\mathscr{E}$ is very poor; and the best performance is for the emotion anger at 64.5%.

This motivates the first experiment, namely to identify the smallest unit of speech (namely τ) that can reliably estimate the emotion in the spoken speech segment and also aid in improving the emotion estimation. We split the speech utterance into smaller segments, as shown in (4.3), for values of τ ranging from 100 to 1800 ms. For a given value of $\tau = \tau_p$, we first segmented the speech utterance (say $s(t)$ annotated

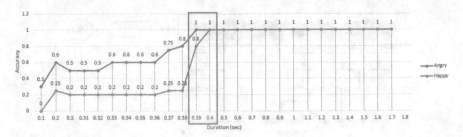

Fig. 4.3 Emotion recognition accuracy as a function of τ for two speech samples (selected at random). Note that a value of $\tau \geq 400$ ms results in all the segments being correctly recognized

as $e_1 = \text{happy}$) into nonoverlapping segments, each of length τ, say $s_1, s_2, \ldots, s_m$. For each of the m segments, we computed the emotion (as in (4.4)). The number of segments which was recognized as happy were counted (say γ; note that $\gamma \leq m$), and a measure γ/m was computed and assigned to τ_p. Figure 4.3 shows the ratio γ/m for different values of τ (x-axis) for two different speech files selected at random from Emo-DB. Clearly, as seen in Fig. 4.3, the performance deteriorates when the value of $\tau < 390$ ms for both the sample audio utterances. This experiment helped us determine $\tau = 400$ ms as the smallest length of the speech sample that resulted in all the segments being correctly recognized.

As part of our second experiments, we demonstrate the performance of the proposed approach to compute emotion. As explained earlier, the proposed approach is based on segmenting the give audio into smaller segments, computing the emotion of each segment followed by fusing the identified emotion of each segment. For each of the 343 audio files, we segmented the audio file into smaller segments of duration 400 ms (see Fig. 4.2). For each segment (using the same set of features and emotion extracting scheme as used in Table 4.2), we identify the emotion. The emotion of the spoken initial utterance is estimated from the individual emotions assigned to the segments of the initial utterance. The emotion assigned to the initial utterance is the one which is assigned to a maximum number of speech segments as explained earlier. The results of the proposed approach is shown in Table 4.2. Clearly, the performance of our approach, namely segmenting the audio into smaller segments and computing the emotion of each segment and then fusing the emotions to assign an emotion to the initial unsegmented audio, performs far better than the conventional approach. Compare Table 4.2 (our approach) with the conventional approach (Table 4.1); there is an overall 33% (49.2% $\rightarrow$ 82.5%) improvement in emotion recognition accuracy. This shows that the segmentation based approach is able to identify the emotion of an utterance much better than the conventional approach for the same set of features ($\mathscr{F}$) and the same emotion recognition scheme ($\mathscr{E}$).

Although here we show the improved emotion recognition with acted speech (Emo-DB), but the framework is applicable for realistic speech. In [17], the use of segmented audio for emotion recognition in call center calls enhanced the emotion recognition accuracy.

Table 4.2   Emotion recognition accuracy (%) for the proposed approach

Emotion	Proposed approach (Fig. 4.2b)	
anger	(111/127)	87.40
happy	(59/71)	83.09
neutral	(68/79)	86.00
sad	(45/66)	68.18
Total	(283/343)	82.50

Table 4.3   Emotion recognition accuracies (%) using different acoustic feature

Train–test set	Feature sets	Accuracy	
		80% train 20% test	10-fold cross validation
Emo-DB	IS09_emotion—(384 features)	74.76	75.88
	Emobase—(988 features)	76.63	79.62
	IS10_paralinguistic—(1582 features)	79.24	82.05
	Emo_large—(6552 features)	81.13	84.67

4.3   Feature Level: Robust Features for Improved Speech Emotion Recognition

Automatic emotion recognition with good accuracies has been demonstrated for noise-free acted speech, which is recorded in controlled studio environment. Extraction of large set of diversified features that closely resemble the exact emotional class information helps to improve the emotion recognition accuracy. Initially, prosodic features were used [42, 53], slowly people started to use large set of features with different characteristics (perceptual, spectral, temporal, spectro-temporal, energy etc.) The advantage of such large feature vector is not only limited to its use in clean environment, but also for speech which is contaminated by noise [100]. The large set of feature extraction technique is depicted in Fig. 4.4. The idea here is to extract large number of statistical features (also known as the high-level descriptors) from the low-level descriptors for the purpose of emotion recognition from speech; no matter whether the speech is clean or noisy. Then, feature reduction and selection techniques are adopted to reduce and find the relevant features that are useful for classification (Table 4.3).

Experiments are conducted with four different set of features using the feature configuration files available in openSMILE toolkit. In first set of experiments, we divided the Emo-DB dataset into 80% train and 20% test. And in the second set, we conducted experiments using 10-fold cross validation. The best result is obtained with the emo_large configuration file and with 10-fold cross validation. The large set of diversified features also inspired us to check its robustness in noisy space as

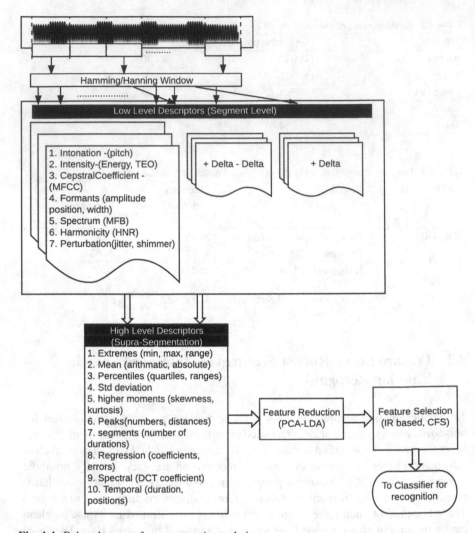

Fig. 4.4 Robust large-set feature extraction technique

well. Table 4.4 presents the affective content detection accuracies for the acted speech samples (from Emo-DB dataset), which are contaminated by different levels (SNR ranging from −5 dB to 20 dB) of four different types of noise (babble, F-16, machine-gun, and Volvo) from Noisex-92 dataset. The noise was added using FaNT Toolkit [30]. Noise contaminated acted utterances are segmented to 400 ms smaller segments like we did in [73]. We have used three different classifiers, namely SVM, ANN, and k-NN and fused their output using product-rule-based classifier combination [45]. As expected for lower SNRs, the accuracies are on the lower side and improved with higher SNRs. Combining classifier scores improves the accuracy. It is also observed that the performance of the system degrades significantly when the signal

Table 4.4 Detection accuracies for acted speech (Emo-DB) contaminated by noise (Noisex-92)

Noise type	SNR (dB)	Classifiers			
		SVM	ANN	k-NN	SVM+ANN+k-NN
Babble	−5	21.05	22.3	32.6	33.2
	0	22.1	22.8	32.9	33.9
	5	24.8	25.6	34.6	35.5
	10	28.9	30.1	35.2	37.2
	20	42.3	45.6	47.3	53.3
F-16	−5	20.8	20.8	28.3	30.6
	0	21.6	22.3	29.1	32.4
	5	22.7	23.8	30.5	39.6
	10	28.3	30.1	34.6	43.7
	20	30.7	33.4	38.6	45.2
Machine-gun	−5	22.8	34.6	41.1	47.1
	0	45.6	61.4	61.4	73.3
	5	70.8	71.2	63.2	75.4
	10	71.9	73.2	64.9	77.2
	20	72.3	76	68	80.2
Volvo	−5	20.3	22.8	32.9	37.3
	0	40.3	42.7	40.3	52.9
	5	49.1	49.8	42.1	57.6
	10	54.3	54.8	57.3	71.3
	20	66.7	66.9	72.7	77.3

is affected by babble noise, and comparatively, lower degradation is observed with the machine-gun noise.

4.4 Improvement at Classifier Level Using Error Correcting Codes

In this section, we propose to use error correction coding (ECC) for improving the emotion identification accuracy. We conceptualize the emotion recognition system is a noise-affected communication channel, which motivates us to use ECC in the emotion identification process. It is assumed that the emotion recognition process consists of a speech feature extraction block followed by an artificial neural network (ANN) for emotion identification, where emotions are represented as binary strings. In our visualization, the noisy communication channel is "not sufficiently" learnt ANN classifier that in turn results in an erroneous emotion identification. We apply ECC coding to encode the binary string representing the emotion class by using a block coding (BC) technique. We experimentally validated that applying ECC coding and decoding improves the classification accuracy of the system in comparison to the baseline emotion recognition system based on ANN. Our initial discussion shows that ECC+ANN performs better than the only the ANN classifier, justifying the use of ECC+ANN. The conjecture is also made that the ECC in ECC+ANN can be visualized as a part of deep neural network (DNN). It is also shown that the use of ECC+ANN combination is equivalent to the DNN, in terms of the improved classification accuracies over the ANN.

4.4.1   Relevance of ECC in Emotion Recognition

Error detection and correction codes are used for the reliable transmission of the data over a transmission channel [31]. In practice, information media are not reliable perfectly, in the sense that interfering noise most often distorts the data. Data redundancy is incorporated in the original data to deal with such adverse situation. With such redundancies, even if errors are introduced, original information can be recovered. ECC adds redundancy to the original information so that at the receiver, it is possible to detect and correct the errors and thus recover the original message [31, 62]. It is helpful for those applications where resending of the message information is not feasible, just as in case of an ANN classifier.

Human Computer Interactive (HCI) system can be benefited by addressing paralinguistic cues such as human emotions (e.g., angry, happy, neutral), body postures, gestures, and facial expressions. Interestingly, emotional state of human has a direct bearing on the characteristics of spoken utterances [4, 71]. Therefore, human emotion recognition can improve the performance of HCI systems [21]. Conventional speech emotion recognition systems extract audio features like pitch, formants, short-term energy, mel-frequency cepstral coefficients (MFCC), and Teager energy operator-based features prior to recognition [3, 35, 67, 95, 113] and then combined followed by feature selection method [95, 96]. Once the features are extracted, the focus shifts to choosing a learning algorithm. Several classifiers have been used in speech emotion recognition, including SVM, ANN, HMM, GMM, Bayesian Network (BN), k-nearest neighbor (k-NN).

We discuss, how to improve the identification accuracies using an ANN classifier to learn different emotions and propose a system that corrects the output of the ANN. We visualize the ANN as a kind of transmission channel problem where the identity of the actual output emotion class for unseen sample is being transmitted over the channel [13, 14]. The channel here is analogous to the "input features," "training examples," "learning algorithm." Because of errors, which may be introduced by the choice of input features, the insufficient training samples and the learning process, the emotional class information is corrupted. It is hypothesized that by encoding the class using ECC and transmitting each bit separately, the system may be able to correct the errors. We hypothesize that a DNN (with nodes in the multiple hidden layers greater than the nodes in the input layer) is equivalent to the ECC-ANN

Table 4.5 Emotions represented in terms of 7 bits

o_1	o_2	o_3	o_4	o_5	o_6	o_7	Emotion
1	0	0	0	0	0	0	anger
0	1	0	0	0	0	0	boredom
0	0	1	0	0	0	0	disgust
0	0	0	1	0	0	0	fear
0	0	0	0	1	0	0	happy
0	0	0	0	0	1	0	neutral
0	0	0	0	0	0	1	sad

combination. Our hypothesis is based on the idea that a DNN with multiple hidden layers can capture more complex representations from the input, thus allows a deep network to have significantly higher representational power, i.e., can learn complex functions (conceptually similar to the ECC) than an ANN. We visualize the ECC in ECC-ANN combination as a part of a DNN [14].

4.4.2 Problem Formulation

We considered seven categorical emotion classes, namely anger, boredom, disgust, fear, happy, neutral, and sad. And those classes are encoded with the binary bit stream, lets say, "1 0 0 0 0 0 0" presenting anger, "0 1 0 0 0 0 0" represents boredom, "0 0 1 0 0 0 0" represents disgust, "0 0 0 1 0 0 0" presenting fear, "0 0 0 0 1 0 0" represents happy, "0 0 0 0 0 1 0" presenting neutral, and "0 0 0 0 0 0 1" presenting sad. Therefore, emotion recognition problem is the following: "given an audio signal, to classify it into one of the seven classes (anger, boredom, disgust, fear, happy, neutral, and sad)."

For the purpose of demonstration, we use the ANN as the classifier that is trained using train dataset. For motivation, we assumed that the ANN to be a transmission channel which takes the speech features as input and transmits a seven-bit output. Like any non-ideal transmission channel, there is a channel transmission error which results in output bits in error. It is well known that ECC is used for handling channel transmission errors in a communication system. It has to be noted that in our case, the channel is the ANN, which might give out erroneous output. This observation motivates us to use ECC in the emotion recognition process.

4.4.3 Speech Emotion Recognition System

Let us consider $\{\mathscr{F}_{x_i}^j\}_{j=1}^J$ be a set of J audio features extracted from the audio clip $x_i(t)$ and assume that we have seven classes of emotions, namely, anger, boredom, disgust, fear, happy, neutral, sad. Let each emotion be represented by $k = 7$ bits, namely $o_1, o_2, o_3, o_4, o_5, o_6$, and o_7. A typical ANN-based speech emotion recognition system would take as input $\{\mathscr{F}_{y_i}^j\}_{j=1}^J$ the speech features extracted from the test speech utterance $y_i(t)$ and produces a k-bit output. The k-bit output determines the emotion of the speech utterance $y_i(t)$. It has to be noted that the ANN output could take any of the 2^7 binary states. However, the emotion is deemed recognized if and only if it is one of the seven-bit sequence shown in Table 4.5. Here, the ANN has been trained using an annotated training dataset, namely

$$\{\{\mathscr{F}_{x_i}^j\}_{j=1}^J; o_1^i, o_2^i, o_3^i, o_4^i, o_5^i, o_6^i, o_7^i\}_{i=1}^S$$

where S is the number of training samples.

It is hypothesized that the performance of an ANN-based emotion recognition system can be improved by introducing the ECC concept (as shown in Fig. 4.5). n bit binary string is constructed from seven bits $o_1, o_2, o_3, o_4, o_5, o_6, o_7$ using ECC coding (described in Sect. 4.4.3.1). We construct $e_1, e_2, \ldots e_n$ from $o_1, o_2, o_3, o_4, o_5, o_6, o_7$ (namely by adding $(n-k)$ parity check bits) (as shown in Fig. 4.5). Mathematically,

$$(e_1, e_2, \ldots e_n) = ECC(o_1, o_2, o_3, o_4, o_5, o_6, o_7) \tag{4.5}$$

We have the following training set of S sample to train an ANN

$$\{\{\mathscr{F}_{x_i}^j\}_{j=1}^J, e_1^i, e_2^i, \ldots, e_n^i\}_{i=1}^S$$

For testing, we use the features $\{\mathscr{F}_{y_i}^j\}_{j=1}^J$ extracted from the test speech utterance $y_i(t)$ as the input to get a n bit output, namely $e_1, e_2, \ldots, e_n$. Then, we use ECC decoding to get back $o_1, o_2, o_3, o_4, o_5, o_6, o_7$, namely

$$(o_1, o_2, o_3, o_4, o_5, o_6, o_7) = ECC^{-1}(e_1, e_2, \ldots e_n) \tag{4.6}$$

The $y_i(t)$ is classified as one of the seven emotions if $o_1, o_2, \ldots, \ldots o_7$ take one of the values in Table 4.5. ECC^{-1} is a decoding scheme described in the following Sect. 4.4.3.1.

We also try to visualize ECC and ECC^{-1} as being equivalent to a DNN architecture (see Fig. 4.6). In some sense, if the number of hidden nodes in the first hidden layer "n_{h_1}" is more compared to the number of input nodes "n_i", then the architecture $n_i \rightarrow n_{h_1}$ is equivalent to the k bit emotion being encoded as n bit using the theory of ECC ("Stage 1" in Fig. 4.6). And, if the number of nodes "n_{h_L}" in the last hidden layer is greater compared to the number of nodes in the output layer "n_o," then the

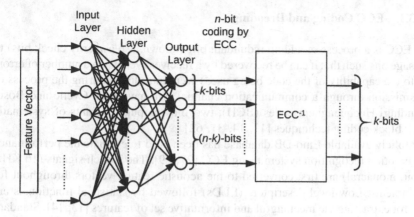

Fig. 4.5 ECC-ANN framework for speech emotion recognition

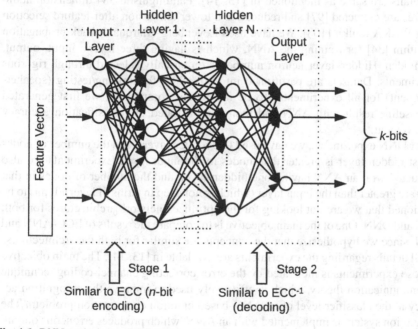

Fig. 4.6 DNN equivalent to ECC-ANN combination

architecture $n_{h_L} \rightarrow n_o$ is equivalent to the decoding (ECC_i^{-1}) of n bit emotion to the k bit using ECC theory ("Stage 2" of Fig. 4.6).

4.4.3.1   ECC Coding and Decoding

The ECC is a process of adding redundant bits (known also as parity check bits) to message bits such that it can be recovered by a receiver even when a number of errors (up to the capability of the code being used) were introduced during the process of transmission through a communication channel. We have used Cyclic and Bose-Chaudhuri-Hocquenghem codes (BCH), two most popular variants of systematic linear block coding techniques [13, 14, 31, 62].

Publicly available Emo-DB database has been used to evaluate the performance of a emotion recognition system using ECC [27, 29]. The speech signals (16 KHz, 16 bit, monaural) are first converted to the acoustic feature vectors throughout for experiments. Low-level descriptors (LLDs) followed by statistical functionals are used for extracting the meaningful and informative set of features [13, 14]. Standard toolkit openSMILE has been used for speech feature extraction [70]. The feature set details are same as mentioned in [13, 14]. Paralinguistic 384 dimension audio features are extracted [97] and reduced to a lower dimension after feature selection using WEKA toolkit [108]. We have used Levenberg–Marquardt back propagation algorithm [34] for training the ANN, which is having three layers: input, output, and hidden. Hidden layer neuron number is empirically selected through rigorous experiments. Datasets are partitioned into 80%/20% for training/testing (speaker-dependent) for all experiments. To be fair in our comparison, we first generated the baseline result with ANN and then demonstrated the classification accuracy improvement.

For DNN experiments, we have used two hidden layers, and the number of nodes in first hidden layer is greater than nodes in the input layer. Experiments are also conducted with an ANN having one hidden layer, and the number of nodes in that layer are greater than the input layer (which is not deep in terms of layer). It has to be mentioned that we are not looking for the optimized network architectures for both ANN and DNN. One of the main objective is to compare the results of ECC-ANN and DNN, since we hypothesize that the concepts are similar for both the architectures.

All details regarding the experiments are available in [13, 14]. The main objective of those experiments is motivated by the error correcting channel-coding technique in communication theory, which is effectively used to improve the recognition accuracy at the classifier level of an ANN-based emotion classification problem. The recognition system is implemented with an ANN, which produces erroneous output, just like a communication channel introduces errors in the transmitted information. Experiments are performed by encoding the emotion classes in categorical space with two popular error correcting linear block codes. Improvement in the recognition accuracies due to the use of ECC at the output of a discriminative ANN classifier favorably shows the use of ECC in speech emotion recognition. Moreover, experimental results also validate the improvement in error correction capability with a systematic increment in the codeword length, but with an increased time complexity. Moreover, the experimental result also validates our hypothesis that a deep neural network architecture is equivalent to the ECC-ANN combination. In other words, we can say that ECC in an ECC-ANN combination framework can be considered as a part of DNN where the intelligence is under control. The enhanced performance of ECC-ANN can be attributed to the larger distance between the emotions in elongated codewords (which are systematically found through ECC). Moreover, DNN with multiple hidden layers has significantly greater representational power, which can learn significantly complex functions and performs similarly to the ECC-ANN combination. Interestingly, experimental result validates that the ECC-ANN combination performs similarly to the DNN.

4.5 Conclusion

Recognition of emotion in spontaneous speech is a hard problem with several challenges. In this chapter we have seen three methods of overcoming them. The first one at the signal level by segmenting audio, the second one at the feature level by the choice of suitable and robust features, and the third one by using ECC at post classifier (ANN) level. We show the improvements in emotion recognition accuracy by using any one of these three methods.

Chapter 5
Case Studies

5.1 Mining Conversations with Similar Emotions

Detecting emotions in a large volume of Call Center calls is essential to identify conversations that require further action. Most emotion detection systems use statistical models generated from the annotated emotional speech. However, annotation is a difficult process that requires substantial amount of efforts, in terms of time and cost. Also human annotation may not be available for call center calls because of the infrastructure issues. Recognition systems based on readily available annotated emotional speech datasets (mostly recorded in clean environment) give erroneous output due to train–test condition mismatch for realistic speech. Therefore, the idea here is to design a framework that can automatically identify similar affective spoken utterances in a large volume of call center calls by using emotion models which are trained with the available acted emotional speech. As discussed in Chap. 3, to reliably detect the emotional content, we incorporate all available knowledge associated with the call (namely time-lapse of the utterances in a call, the contextual information derived from the linguistic contents, and speaker information). For each speech utterance, the emotion recognition system generates similarity measures (i.e., likelihood scores) in `arousal` and `valence` dimension. These scores are further combined with the scores from the contextual knowledge-based modules.

5.1.1 Introduction

With the increasing number of call centers in recent days, affective content[1] analysis of audio calls is important [43, 52, 77]. Manual analysis of these calls is cumbersome and may not be feasible because large volume of recordings happen on daily basis. So, only a small fraction of these conversations is carefully heard by human

---

[1]Affective content and Emotion will be used interchangeably in this chapter.

© Springer Nature Singapore Pte Ltd. 2017
R. Chakraborty et al., *Analyzing Emotion in Spontaneous Speech*,
https://doi.org/10.1007/978-981-10-7674-9_5

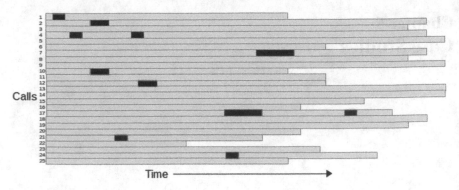

Calls

Time ────────────────────────────────────►

Fig. 5.1 Call center calls (dark portions indicate problems)

supervisors and addressed, thus resulting many of those unattended. This results in many problematic calls going unattended.

Figure 5.1 illustrates the difficulty of identifying the affective (or emotionally rich) regions manually in large number of calls. The call duration is plotted on the x-axis, while different calls are shown along the y-axis. Note that the calls are of different durations and the gray color represents the actual length of the calls. The black color within the call shows the location of an affective state (highly correlated to the problematic regions in the calls). However, the location of problematic regions are arbitrary and the durations are of variable length, making it more challenging for machine to analyze. In spite of this, affective analysis of call center conversations has attracted the attention of researchers (e.g., [33, 43, 60, 79, 82, 111]).

Affective content analysis is a technique that extracts emotions from spoken utterance.[2] In general, affective contents are represented categorically in terms of emotion classes (e.g., [33, 79, 82, 111]). Four emotions (namely anger, happy, neutral, sad) are important in call center calls. Capturing time-varying emotions in dimensional space using audio-visual cues has been proposed in [66], while an incremental emotion recognition system that updates the recognized emotion is proposed in [60]. Combining linguistic information with the acoustic features found to be beneficial in terms of the performance. In [48], authors proposed a combination of acoustic, lexical, and discourse information for emotion recognition in spoken dialogue system and found improvements in recognition performance. Similarly in [87], authors described an approach to improve emotion recognition in spontaneous children's speech by combining acoustic and linguistic features.

In this case study, we have used a novel framework that automatically extracts the affective content of the call center spoken utterances in arousal and valence dimensions [16]. We also fuse context-based knowledge (e.g., time-lapse of the utterances in the call, events and affective context derived from linguistic content, and speaker information) associated with the calls. For each spoken utterance, the affective content extractor generates probability scores in arousal and valence

---

[2]The word utterance, turn, and spoken terms will be used interchangeably from here on.

dimensions, which are then probabilistically combined to label it with any of the predefined affective classes. In this framework, emotions are extracted at discrete levels of affective classes (i.e., `positive`, `neutral`, `negative` in `arousal` and `valence` dimensions), instead of using affective information in continuous scale like in [66]. This reduces the complexities related to the difficulties in annotating affective states on a continuous scale. The reduced number of classes in each dimension makes it easy to annotate. In addition, detection of similar emotions in the calls are performed by using the emotion models trained with acted emotional speech (Emo-DB). The system can work even in a scenario when annotated call center calls are not available.

5.1.2 Motivation and Challenges

Voice-based call center records thousands of calls on a daily basis. It is difficult for supervisor to identify emotional segments among these large number of calls to decide which of these calls need actions to address customers dissatisfaction. In general, the calls are of variable duration, varying from a few seconds to couple of minutes depending on factors like the type of the call center, type of the problems that the customers are facing, the prior affective state of customer, the way agent handles the problem, etc. In this scenario, a super agent (or supervisors) manually selects randomly, few calls from large number of calls to listen and check if there are some abnormalities. Therefore very few calls are normally analyzed and addressed, thus resulting in many calls going unattended. This often leads to an increasing dissatisfaction among the customers. Automation of this process is useful to deal with such situation. Most often statistical models generated from available annotated data are used to detect the affective states in audio calls.

We design a system that learns different affective states from available annotated emotional speech. Since, there is a mismatch in train-test scenario, it produces erroneous output. To handle this mismatch condition, our system intelligently makes use of available knowledge for reliable extraction of affective state. For an audio segment in a call, the system generates a probability matrix whose elements are a joint probability estimate in the affective dimension of `arousal` and `valence`. Then we fuse the knowledge-based information to modify the elements of the matrix. Similarity is measured by calculating the distances between the modified matrix and the reference matrices. In this way, similar segments are bucketed together and it becomes easier for the supervisors to quickly identify and analyze all problematic calls depending on the labeled affective content.

5.1.3 Mining Similar Audio Segments

The framework for mining similar audio segments consists of several blocks (see Fig. 5.2). The blocks are (i) affective content extractor that gives scores for different affective states in the two-dimensional (`arousal` (A) and `valence` (V)) space, (ii) three knowledge-based modules (knowledge regarding audio segmentation, time-lapse of the segment in the call, and voice to text analytics), (iii) decision block for deciding similar segments. When an audio call is fed to the framework for analysis, it is passed through a speaker segmentation system which segments an audio call into different segments of agent's and customer's voice. Call center calls, generally consists of voice of two speakers (agent and customer). As depicted, depending on the requirement, the framework is able to pick up all the segments associated with the agent or associated with the customers by selecting appropriate value of (w_A, w_c) pairs (see Fig. 5.3). The system chooses (1, 0) for agent's audio only and (0, 1) for customer's audio only.

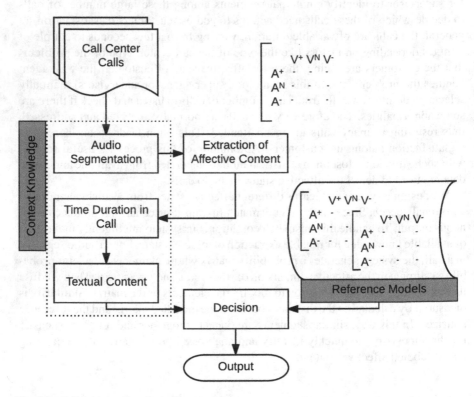

Fig. 5.2 Call center call analysis framework

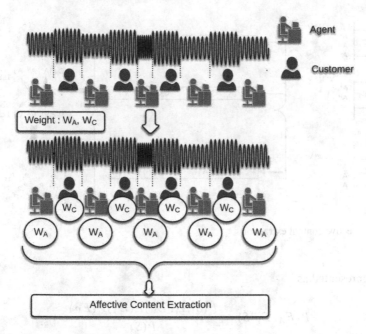

Fig. 5.3 Audio segmentation-based knowledge

5.1.4 Affective Content Extraction

Let us assume that we have annotated data of a large corpus in two dimensions, namely A and V. Further, let there be three classes in each affective dimension, i.e., $E_A = \{A^+, A^N, A^-\}$ and $E_V = \{V^+, V^N, V^-\}$. Let us assume that we have statistical models for A^+, A^N, A^-, V^+, V^N, and V^- such that we can compute for an audio segment S the following, $P(E_A|S)$ (i.e., $P(A^+|S)$, $P(A^N|S)$, and $P(A^-|S)$) and $P(E_V|S)$ (namely, $P(V^+|S), P(V^N|S)$, and $P(V^-|S)$). Therefore, we represent an audio segment S in the A-V space by a 3×3 matrix at the output of each classifier $(c_1, c_2, ..., c_C)$, namely

$$P(E|S) = \begin{bmatrix} P(A^+|S)P(V^+|S) & P(A^+|S)P(V^N|S) & P(A^+|S)P(V^-|S) \\ P(A^N|S)P(V^+|S) & P(A^N|S)P(V^N|S) & P(A^N|S)P(V^-|S) \\ P(A^-|S)P(V^+|S) & P(A^-|S)P(V^N|S) & P(A^-|S)P(V^-|S) \end{bmatrix}.$$

Affective content of a segment S is defined by

$$\varepsilon^k_{A,V} = P(E|S) = P(E_A, E_V|S) \tag{5.1}$$

where $\varepsilon^k_{A,V} = P(E_A, E_V|S)$ is the posterior score associated with S being labeled as emotion E_A and E_V, using a trained emotion recognition system. The posterior

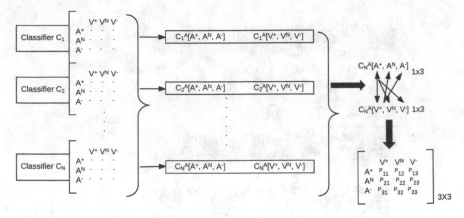

Fig. 5.4 Affective content extraction

can be represented as

$$P(E_A, E_V | S) = \frac{P(S|E_A, E_V) P(E_A) P(E_V)}{P(S)}$$

where $P(S|E_A, E_V)$ is the likelihood, $P(E_A)$ and $P(E_V)$ are the priors. Assuming that affective contents at the `arousal` and `valence` dimensions are independent, we can write

$$P(S|E_A, E_V) \approx P(S|E_A) P(S|E_V)$$

Note that S is defined as $\chi(x(\tau - \Delta\tau), x(\tau))$, where χ is the operator that extracts high-level features from the audio signal between the time interval $(\tau - \Delta\tau)$ and τ. High-level features are typically statistical functionals and are constructed from low-level descriptors, which operate in the interval of $\tau - \Delta\tau$ and τ of the signal x.

From the output of each classifier, we construct a 3×3 matrix, whose elements are posterior probabilities ($\varepsilon_{A,V}^k$) (as shown in Fig. 5.4). According to (5.1), $\varepsilon_{A,V}^k$ can have two set of elements, namely, ε_A^k and ε_V^k. As an example, lets say for the first classifier, we have the scores ε_{A+}^1, $\varepsilon_{A^N}^1$, and ε_{A-}^1 for the emotion in `arousal` dimension. Similarly for the same classifier, we have the scores ε_{V+}^1, $\varepsilon_{V^N}^1$, and ε_{V-}^1 in the `valence` dimension. Then $\varepsilon_{A,V}^k$ (3×3 matrix) is obtained considering each pair from ε_A^k and ε_V^k. Then the recognition system outputs a posterior probability matrix for the utterance $x(\Delta\tau)$ by combining the scores from all the classifiers, which is given by

$$\varepsilon^{ER}(x(\Delta\tau)) = \sum_{k=1}^{C} \varepsilon_{A,V}^k \tag{5.2}$$

where $\varepsilon^k \in \mathcal{E}_{A,V}$ is the estimated joint probability scores of the utterance $(x(\tau - \Delta\tau), x(\tau))$. This works well with acted speech where one has the luxury of annotated training data (i.e., $(x(\Delta\tau), E_A E_V)$ pairs) to build several classifiers. However, with the spontaneous speech like call center calls, probability estimation may be erroneous because of the mismatch conditions. However, the estimations can be improved by using knowledge.

5.1.5 Using Time-Lapse Knowledge of the Audio Segment

The probability matrices that we get from affective content extractor are passed through the knowledge-based module, which modifies the probability scores depending upon the time-lapse of the segment in the audio call (as shown in Fig. 5.5). We observe through analysis that the total duration of audio call plays an important role in influencing (or changing) the user's affective state. In such scenario, we hypothesize that the intensity of some of the affective states (namely A^+, V^-) increases and some (namely A^-, V^+) decreases with time. The output can be represented as

$$\varepsilon_i^{lapse} = w_i \varepsilon^{ER} \tag{5.3}$$

where ε_i^{lapse} is the output matrix we get after multiplying the weight matrix w_i at the time index i. As shown in the Fig. 5.5, it is expected that weights for (A^+, V^-) affective content at the end of the call will be more in comparison to its weight at the beginning of the call. We hypothesize that these weight increase exponentially as time index i increases for A^+ and V^- and decreases exponentially as time index i increases for A^- and V^+ (see Fig. 5.5).

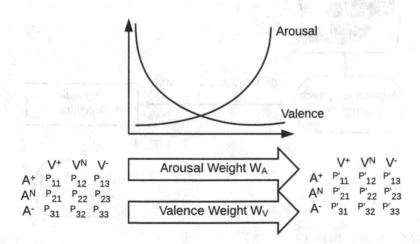

Fig. 5.5 Knowledge regarding the time-lapse of the segment in a call

5.1.6   Speech to Text Analytics

The modified matrices are then passed through linguistic knowledge-based module (as shown in Fig. 5.6), which converts the spoken utterances into text (by using an ASR), followed by the text analytics block to generate a weight matrix w_t that consists of the probabilities of affective state given a spoken word or phrase. To analyze the affective state of customer's voice at any instant of time, the immediate previous spoken words of the agent is considered. The same process is followed for analyzing the agent's voice. It is observed that the spoken words from one speaker at any instant of time induce some specific affective state in the other user during a call conversation. This knowledge-based module consists of two subsystem, (i) an ASR engine and (ii) a voice analytics engine. The ASR engine converts the spoken words into text, and the voice analytics module learns and spots emotionally prominent words so as to improve the overall emotion recognition. We used a prominence measure to find and associate words that are related to affective states in the speech data. With the affective prominence, we created a weight matrix w_t, where each element represents the affective prominence corresponding to an affective state. For each utterance, we have measure of the affective prominence for each affective states, which forms the weight matrix w_t to modify the output of second knowledge-based module as shown in Fig. 5.6.

$$\varepsilon^{vta} = w_t \varepsilon^{lapse} \tag{5.4}$$

where ε^{vta} is the output matrix we get after multiplying weight matrix w_t.

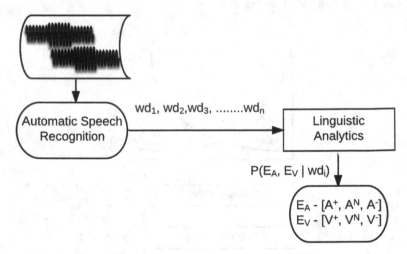

Fig. 5.6 Speech to text analytics

5.1.7 Decision Based on Similarity Measures

We obtain the output matrix ε^{vta} for each audio segment and then find the distance from the reference matrix (i.e., a template for each affective state) to calculate the similarity measure. The distance d between ε^{vta} and $\varepsilon_{A,V}$ is calculated as

$$d = \| \varepsilon^{vta} - \varepsilon_{A,V} \| \tag{5.5}$$

where $\| \cdot \|$ is the Euclidean norm between the ε^{vta} and the reference $\varepsilon_{A,V}$. Now the affective state of the audio segment is hypothesized as the output corresponding to the reference matrix that has the minimum distance from the output matrix ε^{vta}. It is also possible to arrange the segments in the descending order, i.e., from the highest to the lowest distance. Therefore at the output, audio segment is labeled with the affective class, and its distances from all affective classes.

5.1.8 Experimental Validations

5.1.8.1 Database

For experimentations, we considered call center conversations between different agents and customers. It is to be noted that the agents are trained to talk normally in any situation, mostly suppressing their true emotions while talking to the customers. On the other side, customers generally express their emotions while talking to the agents, which means that the customer speech is expected to be spontaneous and non-acted.

We have considered a total of 107 call center calls from three different sectors (37 calls from finance, 34 calls from telecommunication, and 36 calls from insurance). There are a total of 354 randomly selected audio utterances of the customer which are considered for validating our framework. It is to be mentioned that we found an average SNR level of 8.15 dB for call center calls. Notice that each of the spoken utterance has a reference in the form of when in the conversation flow it was spoken, plus also the manual transcription of speech (temporal sequence of words and phrases). We asked 7 human evaluators to annotate the emotion expressed in each of the 354 utterances by assigning it an emotion label from the set of arousal (positive, neutral, and negative) and valence (positive, neutral, and negative). In the first set of experiments, we randomly sequenced the utterances (with some utterances repeated) so that the evaluators had no knowledge of the events preceding the audio utterance and we then asked the evaluators to label the utterances; while in the second set of experiments, we provided the utterances in the order in which they were spoken along with the spoken words transcription. The motivation to include knowledge related to the conversation at the time of annotation is based on our observation that there is significant disagreement among human annotators when they annotate call center speech; however the disagreement largely reduces significantly when they are provided with additional knowledge related to the conversation. We computed the Kappa score β [112] on the annotations in arousal dimension for each of the two settings. In the first set of experiments (Table 5.1), we obtained a score of $\beta = 0.14$ (i.e., without knowledge), suggesting a very poor agreement between the evaluators. While in the second set of experiments, we obtained $\beta = 0.76$ (refer Table 5.1), suggesting that there was an improved degree of agreement between the evaluators (i.e., with the knowledge). This clearly demonstrates that there was a better consistency in the evaluator's annotation when they were equipped with prior information (knowledge) associated with the utterance. This observation form a basis for reliable recognition of affective states in call center speech in our framework.

5.1.8.2   Experimental Results

We considered audio calls which are manually "speaker segmented" so that the speaker segmentation errors do not propagate to the affective content extractor [16]. Similarly for the speech to text conversion, instead of taking the ASR output, we considered the manual transcription of the audio calls. Affective content extraction system is trained with the acted speech utterances from Emo-DB [27]. Since the Emo-DB dataset has annotations in categorical space, we converted the labels into the dimensional space by considering the 2D model of representation of emotion. All the audio samples in our experimentations are sampled at 8 kHz, 16 bit, and monaural.

Different classifiers SVM, artificial neural network (ANN), and k-NN have been used in the experiments. LibSVM toolkit is used for implementing SVM classifier [50]. For ANN, we have used feed-forward multilayer perceptron [108], and the network is trained with back-propagation algorithm. All the results are presented as an average detection accuracies over all classes. Two different approaches were adopted for extracting the features from the speech utterances, (1) considering the full utterances (2) splitting the audio in 400 ms as mentioned in [73]. In the second case, classifier scores are combined to get the scores for the full utterance.

Table 5.2 represents the affective content detection accuracies for the segments using different classifiers plus their combination and also using different knowledge information. It is observed that combining classifier scores using add rule improves the recognition accuracies [45]. The recognition accuracies are improved by using only the knowledge of the segment's lapse in the audio call. Similar trend is observed when only the voice to text analytics knowledge module is used, and the accuracies were better compared to the system when only time-lapse-based knowledge is used. Moreover, the better accuracies are obtained when all available knowledge is incorporated. The best accuracies are obtained with the framework that segments the full utterance into 400 ms smaller segments compared to the system which uses the full utterance for processing. A significant absolute improvement in accuracy of 23.8%

Table 5.1 Kappa statistics of arousal with and without knowledge

	< 0—No agreement 0.0–0.19—Poor 0.20–0.39—Fair 0.40–0.59—Moderate 0.60–0.79—Substantial 0.80–1.00—Perfect	Kappa score—without knowledge						Kappa score—with knowledge (time-lapse + linguistic content)					
No of samples		+ve	N	-ve	Raters	P_i	P_o	+ve	N	-ve	Raters	P_i	P_o
	U1	7	0	0	7	1.00	0.47	7	0	0	7	1.0	0.86
	U2	1	1	5	7	0.48		0	1	6	7	0.71	
	U3	3	0	4	7	0.43		7	0	0	7	1.00	
	U4	2	0	5	7	0.52		2	0	5	7	0.52	
	U5	2	0	5	7	0.52		0	0	7	7	1.00	
	U6	5	0	2	7	0.52		6	0	1	7	0.71	
	U7	4	0	3	7	0.43		0	0	7	7	1.00	
	U8	2	2	3	7	0.24		0	7	0	7	1.00	
	U9	2	1	4	7	0.33		7	0	0	7	1.00	
	U10	2	3	2	7	0.24		2	0	5	7	0.52	
	U11	5	0	2	7	0.52		1	0	6	7	0.71	
	U12	0	5	2	7	0.52		7	0	0	7	1.00	
	U13	1	2	4	7	0.33		0	0	7	7	1.00	
	Total	36	14	41	91	6.10		39	8	44	91	11.19	
	Pj	0.04	0.15	0.45		$K = \frac{(Po-Pe)}{(1-Pe)}$		0.43	0.09	0.48		$K = \frac{(Po-Pe)}{(1-Pe)}$	
	Pe	0.38				K = 0.14		0.43				K = 0.76	

Table 5.2 Affective content detection accuracies (%) for call center calls (full and 400 msec split)

Description	Classifiers							
	SVM		ANN		k-NN		SVM + ANN + k-NN	
	Full	Split	Full	Split	Full	Split	Full	Split
Affective content extractor	32.3	62.7	36.1	63.8	31.3	63.2	39.8	65.8
+Time lapse	49.3	78.3	53.9	78.9	44.2	72.5	59.7	81.9
+ASR and text analytics	56.8	80.9	56.2	82.1	45.1	76.1	61.2	85.2
+Time lapse +ASR and text analytics	65.3	87.6	61.2	88.9	47	85.6	72.1	89.6

is achieved when all the available knowledge is used for the combined classifier, for 400 ms audio.

5.2 Affective Impact of Movies

This work describes the MediaEval 2015 Affective Impact of Movies Task (includes violent scene detection) along with our proposed scheme during our participation in that evaluation. The overall use case scenario of the task is to design a video search system that uses automatic tools to help users find videos that fit their particular mood, age, or preferences [103]. Specifically, there were two subtasks: (1) induced affect detection: the emotional impact of a video or movie can be a strong indicator for search or recommendation; (2) violence detection: detecting `violent` content is an important aspect of filtering video content based on age.

The task requires participants to deploy multimedia features to automatically detect `violent` content and emotional impact of short movie clips. The short video clips are considered as single units for detection purposes with a single judgment per clip. Both tasks use the same videos for training and testing. We proposed to detect the affective impacts and the `violent` content in the video clips using two different classifications methodologies, i.e., Bayesian network approach and artificial neural network approach, which are described in the following subsections. Experiments with different combinations of features make up for the five run submissions.

5.2.1 System Description

5.2.1.1 Artificial Neural Network-Based Valence, Arousal and Violence Detection

This section describes the system that uses ANN for classification. Two different methodologies are employed for the two different subtasks. For both subtasks, the developed systems extract the features from the video shots (including the audio) prior to classification. The proposed system uses different set of features, either from the available feature set (audio, video, and image), which was provided with the MediaEval dataset, or from our own set of extracted audio features. The designed system either uses audio, image, video features separately, or a combination of them. The audio features are extracted using the openSMILE toolkit [70], from the audio extracted from the video shots.

5.2.1.2 Classification

For classification, we have used an ANN that are trained with the samples available for each of those subtask in the development set. As a data imbalance exists for the `violence` detection task (only 4.4% samples are `violent`), for training, we have taken the equivalent number of samples from the two classes. Therefore, we have multiple ANNs, each of them `is` trained with different set of samples. While testing, each sample is fed to all ANN, and then the scores from all ANNs outputs are added using an add rule of combination and deciding as class that has maximum score. A confidence measure is also provided along with the classification. Moreover, while working with the test dataset, the above-mentioned framework is developed for the different set of features. For combining, two different methodologies are adopted. In one, all the scores are added using an add rule before deciding on the detected class. In the second, the best neural network (selection is made while working with development set) is used for each of the different feature sets. Finally, the scores from all the best networks are added and `the decision is made on the maximum score`.

5.2.2 Experiments

The configurations of three different systems (*run*1–*run*3) are the same for the two different subtasks. The first two *run* submissions are based on ANN. In *run*3, the results are obtained with the random guess, based on the distribution of the samples in the development set. In *run*1 for the `violence` detection subtask, 19 different neural networks with `openSMILE` paralinguistic audio features (13 dimensional after feature selection) are trained. In *run*2 for the `violent` subtask, we have trained 19 different neural networks with five different set of features (41 dimensional MediaEval features, 20 dimensional audio MediaEval features, `openSMILE` audio features (7 dimensional after feature selection), `openSMILE` paralinguistic audio features (13 dimensional), and combination of `openSMILE` audio and MediaEval video and image features). So, we have $19 * 5 = 95$ neural networks. The best five neural network classifiers are selected while working on the development set (Fig. 5.7). The development set is partitioned into 80% and 20% for training and testing, respectively. For the affective impact task, as the data imbalance is not up to that extent as

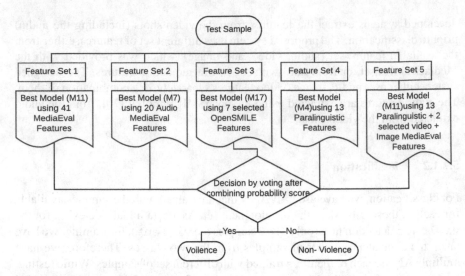

Fig. 5.7 Violence detection using ANN

Table 5.3 MediaEval 2015 results on test set

	Affective impact (accuracies in %)		Violence detection (MAP)
	valence impact	arousal impact	violence
*run*1	33.89	45.29	0.0459
*run*2	29.79	48.95	0.0419
*run*3	33.37	43.97	0.0553

for the `violence` detection, we have trained neural networks `separately` with each of the feature sets for *run*1 and *run*2 (Fig. 5.7).

Table 5.3 shows the results with the metric proposed in MediaEval 2015 [103]. The best result (i.e., 48.95% accuracy) of affective impact detection is obtained with *run*2 for `arousal` detection that combines the best five neural networks that use different feature sets.

Chapter 6
Conclusions

In this monograph, we first established the importance of spontaneous speech emotion recognition and then enumerated the several challenges and hurdles faced during automatic machine recognition of emotion in spontaneous speech. We extensively reviewed the work done in the area of spontaneous speech emotion recognition both in the academic and commercial (patent) space. We enumerated the available databases for emotion recognition research and highlighted the differences between acted, spontaneous and induced emotions. We observed that for spontaneous speech, it is very challenging to (a) generate spontaneous speech database and (b) to obtain robust emotion annotation of the speech database. We brought out these challenges in building an emotion database and outlined a possible approach to build a realistic spontaneous emotional speech corpus [75].

Speech emotion recognition has rich literature for acted speech, which leads to the belief that the techniques for acted speech can be directly used for spontaneous speech. However, there are differences in acoustic characteristics between acted and spontaneous speech which does not allow one to use methods and algorithms that work well for acted speech to recognize emotion in spontaneous speech. Emotion recognition techniques are generally based on machine learning algorithms which require (a) large amount of training speech corpus or database and (b) the test and the train data to have same properties, namely "matched". The main challenge of using trained models that work for acted speech in spontaneous speech is mostly because of the mismatched condition.

The heartline of this monograph is the description of a framework that exploits a priori knowledge to produce reliable spontaneous speech emotion recognition. The main aim of the knowledge-driven framework is to assist the machine learning algorithm with any form of prior knowledge associated with the spontaneous speech. In fact, we hypothesize that the lack of discriminating properties in the audio can be handled by making use of prior knowledge in the form of time lapse of the audio utterances in the call, as well as the use of linguistic contexts. The development of this knowledge-based framework for spontaneous speech emotion recognition is

© Springer Nature Singapore Pte Ltd. 2017
R. Chakraborty et al., *Analyzing Emotion in Spontaneous Speech*,
https://doi.org/10.1007/978-981-10-7674-9_6

both (a) scalable in the sense that it allows for appending any knowledge source and (b) generic in the sense that it boils down to the conventional method of emotion recognition when $\lambda_\kappa = \lambda_l = 0$ (as elaborately explained in Chap. 3).

In this monograph, we also proposed methodologies that can help to improve emotion recognition in spontaneous speech. These methods can be implemented either at the front end (at signal or feature level) or at the back end (at or after the classifier level) of the spontaneous speech emotion recognition framework. Specifically, in Chap. 4, we have discussed three such methodologies. The first one works at the signal level by segmenting the speech signal into smaller speech segments, which are then used for emotion extraction. The second methodology is based on identifying a set of robust speech features, which in turn improve the automatic emotion recognition. And the third methodology described is at the classifier level by using error correction coding (ECC), which are more often used in communication systems. We repurposed ECC by visualizing an emotion recognition system to be a noisy communication channel, which consists of an audio feature extraction module followed by ANN for emotion (represented by a binary string) classification. In particular, ANN is assumed to be an insufficiently learnt classifier which in turn results in an erroneous (binary string) emotion classification. We use ECC to encode the binary string representing the emotion class using a Block Coder (BC) for finding and correcting errors.

From the practitioner perspective, in this monograph, we have discussed two important potential areas where we have used the framework successfully for spontaneous emotion recognition system. We describe two case studies, namely (a) mining conversation with similar affective states in a call center scenario and (b) affective impact of movies on the viewer. In the first case study, automatic detection of emotions in large volume call center audio conversations is essential to spot all those conversations that require further action by the folks handling the call centers. In the second case study, is in line with our participation in the MediaEval 2015 Challenge, we addressed the problem affective impact of movies (includes violent scene detection) on viewers.

Work in the area of automatic recognition of emotions in spontaneous speech is work in progress and demands significant amount of effort to enable its use in real-life problems. As we write this monograph large strides are being made in this area. However, it is important to note that one needs to understand the difference between the spoken speech signal which contains emotion state of the speaker versus the speech signal that inflicts a certain type of emotion on the listener (Appendix B). Today machine learning algorithms are being used without distinguishing the difference in speech signal. It is our belief that, sooner or later, this subtle difference about the speech signal will allow newer techniques to be developed for spontaneous speech emotion recognition.

Appendix A
Emotion Ontology

In the literature, emotion ontology largely discusses different types of emotions. In Fig. A.1, we look at emotion ontology from the perspective of machine recognizing the emotion in an audio.

We can classify emotion based on the category of the emotion (e.g., anger, happy) or based on the affective state, namely as a function of two variables, namely valence (negative or positive) and arousal (high or low). We can also look at emotion being recognized on either acted speech (where a person is asked to act as being in a particular emotional state and then speak a sentence of a phrase), spontaneous speech (this is opposite of acted speech and something that is expressed in day-to-day conversation), or situational speech (this is between acted and spontaneous; the speaker is pushed into a situation without his knowledge and then asked to speak something spontaneous). The coding aspect is covered in more detail in Appendix B.

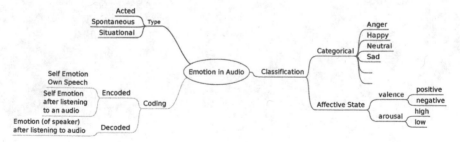

Fig. A.1 Brief emotion ontology for machine learning perspective

© Springer Nature Singapore Pte Ltd. 2017   67
R. Chakraborty et al., *Analyzing Emotion in Spontaneous Speech*,
https://doi.org/10.1007/978-981-10-7674-9

Appendix B
Encoded Versus Decoded Emotion

Primarily most of the work reported on audio emotion recognition does not give due importance to the type of emotion that is being estimated or recognized. Most of them treat the available audio signal as the source of emotion information and then follow the process of extracting features from the audio signal and go about using it as the *data* to be learnt, be it a classifier or a more sophisticated machine learning algorithm.

However, there are very subtle yet important differences that require the reader's attention. Figure B.1 depicts a scenario. Let $a(t)$ be the audio associated with a movie which when seen (or heard) by a speaker can influence his emotion.

- **Scenario 1**: Let us suppose the speaker labels the audio $a(t)$ after he listens to the audio $a(t)$. Call it $E(a(t))$.
- **Scenario 2**: The speaker is in some emotional state, and he speaks, say $x(t)$. Let us say the emotion of the $x(t)$ is $E(x(t))$.

In both these cases, the emotion is labeled by the speaker; however, the first scenario depicts his emotion after listening to $a(t)$, while in the second scenario it is the emotion expressed by him in his speech $x(t)$. Clearly, these two are different.

However, in the literature both these $E(a(t))$ and $E(x(t))$ are treated as being the same in terms of extracting the features from $a(t)$ or $x(t)$ and going about building a machine learning system to identify the emotion. Note that while $x(t)$ has emotion information embedded in the signal itself, the audio signal $a(t)$ does not have the emotion information embedded in itself in the same sense as $x(t)$. Note that $a(t)$ is an input to the speaker (say $H(t)$) who in turn feels the emotion. In fact $E(a(t))$ is $E((H(t) * a(t))$ where $H(t)$ is a transfer function that represents a human which is different for different humans and is influenced by the surrounding environmental conditions.

Also, observe that the reference frame is not the same; in case of $E(a(t))$ or $E(H(t) * a(t))$ the signal is external to the human, while in case of $E(x(t))$ the signal is in the same reference frame as the human. Something similar happens when the signal $x(t)$ is heard by a listener and who in turn assigned an emotion label, say $E(x'(t))$. $E(x'(t))$ is the decoded emotion associated with the signal $x'(t)$, while

© Springer Nature Singapore Pte Ltd. 2017
R. Chakraborty et al., *Analyzing Emotion in Spontaneous Speech*,
https://doi.org/10.1007/978-981-10-7674-9

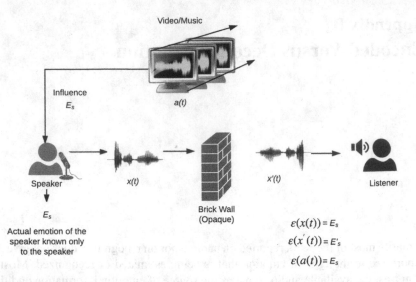

Fig. B.1 Encoded versus decoded emotions

$E(x(t))$ is the encoded emotion associated with $x(t)$. Encoded emotion is the true emotion, while $E(x'(t))$ the decoded emotion is an interpretation of the emotion expressed in $x(t)$. The interpretation of the same signal $x'(t)$ by different listeners could be different, thereby making the decoded emotion very noisy.

So, there are actually three different types of problems that people are trying to solve in the emotion recognition literature without explicitly highlighting this reference frame aspect. Namely,

- $E(x(t))$—encoded, same reference frame, true, example Emo-DB
- $E(x'(t))$—decoded, different reference frame, guess, IVR- SERES
- $E(a(t))$—encoded, different reference frame, true, example MediaEval

In the literature, $x(t)$, $x'(t)$ or $a(t)$ has been treated without distinction as the signal containing the emotion information which is far from being true.

Appendix C
Emotion Databases

In order to develop an acoustic emotion identification system, the foremost requirement is to have sufficient data that extends across the variety and range of affective expressions. As we discussed, earlier in this monograph, spontaneous emotional speech are difficult to collect because of the relative rarity, ethical issues, and complexities related to annotation. Table C.1 briefly describes the prominent and most popular databases which have been used in the field of speech emotion recognition for the last two decades. A detailed description is available at [68].

Table C.1 Select emotion databases (Typ: Acted (A), Induced (I), Spontaneous (S); Data: Audio(A), Audiovisual (AV); Use: Commercial (C), Noncommercial (NC); Lang: English (en), German(gr), Dutch (du), Danish (da), Multilanguage (ML), French (fr); Subjects: Male (M), Female (F)). Comprehensive list at [19]

Year	Name	No of subjects	Duration	Lang	Typ	Data	Usage
2000	SALAS	20 subjects	–	en	I	AV	NC
2002	LDC: Emotional prosody speech and transcripts	30 subjects	9 hrs	en	A	A	C
2003	SMARTKOM: Bavarian archive for speech signals multimodal corpora	224 speakers	130 recording sessions, 490 GB	gr	I	AV	NC
2004	2000 Communicator evaluation DARPA	62 hrs of audio used for speech and emotion recognition	62 hrs and 648 dialogues	en	I	A	C

(continued)

© Springer Nature Singapore Pte Ltd. 2017
R. Chakraborty et al., *Analyzing Emotion in Spontaneous Speech*,
https://doi.org/10.1007/978-981-10-7674-9

Table C.1 (continued)

Year	Name	No of subjects	Duration	Lang	Typ	Data	Usage
2005	eNTERFACE	2 subjects(1 M and 1 F)	150 to 190 utterances for each 6 emotions in 4 languages	ML	A	AV	NC
2007	HUMAINE database	50 recordings	3 s–5 min	en	S	AV	NC
2007	BINED (Belfast Induced Natural Emotion Database)	125 subjects (31 M and 94 F)	10 to 60 s conceptual	en	I	AV	C
2008	IEMOCAP: Interactive Emotional Dyadic Motion Capture	5 M and 5 F	12 hrs	en	A	AV	NC
2009	SEMAINE	20 subjects	–	en	I	A	NC
2009	AviD : Audiovisual speaker identification and emotion detection for secure communications	292 subjects	340 videos	en	A	AV	NC
2010	GEMEP: Geneva Multimodal Emotional Portrayal	10 subjects	–	en	A	AV	NC
2011	RML emotion database	720 audiovisual	4.2 GB	ML	A	AV	NC
2012	RAVDEES: The Ryerson AV Database of Emotional Speech and Song	24 subjects (12 M and 12 F)	7,356 high-quality video recordings	en	A	AV	NC
2015	RECOLA : Remote Collaborative and Affective Interactions	46 speakers	9.5 hrs	fr	S	AV	NC
2015	LIRIS— ACCEDE : A video database for affective content analysis	160 films and short films with different genres are used and segmented into 9800 video clips	–	ML	S	AV	NC

References

1. AWAZYP-CC, Non-Linguistic Analysis of Call Center Call Conversations (2014), https://sites.google.com/site/sunilkopparapu/Home/nonlinguistic-analysis-of-call-center-conversation
2. AWAZYP-SERES, Speech Enabled Railway Enquiry System (2014), https://www.youtube.com/watch?v=fSaNk8CZtGY&feature=youtu.be
3. M. El Ayadi, M.S. Kamel, F. Karray, Survey on speech emotion recognition: features, classification schemes, and databases. Pattern Recogn. **44**, 572–587 (2011)
4. J.-A. Bachorowski, Vocal expression and perception of emotion. Curr. Dir. Psychol. Sci. **8**(2), 53–57 (1999)
5. T. Bänziger, K.R. Scherer, Introducing the geneva multimodal emotion portrayal (gemep) corpus, in *Blueprint for Affective Computing: A Sourcebook* (2010), pp. 271–294
6. BAS, Bavarian Archive for Speech Signals Smartkom—Skaudio 1.0. (2003), http://www.phonetik.uni-muenchen.de/Bas/BasSmartKomAudioeng.html
7. A. Batliner, K. Fischer, R. Huber, J. Spilker, E. Nth, How to find trouble in communication. Speech Commun. **40**(1), 117–143 (2003)
8. I.M. Bennett, Emotion detection device and method for use in distributed systems, 3 July 2012. U.S. Patent No. 8,214,214 (2012)
9. F. Burkhardt, T. Polzehl, J. Stegmann, F. Metze, R. Huber, Detecting real life anger, in *ICASSP* (2009)
10. F. Burkhardt, "you seem aggressive!" monitoring anger in a practical application, in *Proceedings of the Eighth International Conference on Language Resources and Evaluation, LREC 2012*, Istanbul, Turkey, 23-25 May 2012, ed. by N. Calzolari, K. Choukri, T. Declerck, M. Ugur Dogan, B. Maegaard, J. Mariani, J. Odijk, S. Piperidis (European Language Resources Association (ELRA), 2012), pp. 1221–1225
11. C. Busso, M. Bulut, C.-C. Lee, A. Kazemzadeh, E. Mower, S. Kim, J.N Chang, S. Lee, S.S. Narayanan, Iemocap: interactive emotional dyadic motion capture database. Lang. Resour. Eval. **42**(4), 335 (2008)
12. C. Busso, S. Narayanan, Recording audio-visual emotional databases from actors: a closer look, in *Second International Workshop on Emotion: Corpora for Research on Emotion and Affect, International Conference on Language Resources and Evaluation (LREC 2008)* (2008), pp. 17–22
13. R. Chakraborty, S.K. Kopparapu, Improved speech emotion recognition using error correcting codes. In *2016 IEEE International Conference on Multimedia & Expo Workshops, ICME Workshops 2016, Seattle, WA, USA, 11–15 July 2016* (IEEE Computer Society, 2016) pp. 1–6

© Springer Nature Singapore Pte Ltd. 2017

R. Chakraborty et al., *Analyzing Emotion in Spontaneous Speech*,

https://doi.org/10.1007/978-981-10-7674-9

14. R. Chakraborty, S.K. Kopparapu, Validating "is ECC-ANN combination equivalent to dnn?" for speech emotion recognition, in *2016 IEEE International Conference on Systems, Man, and Cybernetics, SMC 2016, Budapest, Hungary, 9–12 Oct* (2016), pp. 4311–4316

15. R. Chakraborty, M. Pandharipande, S.K. Kopparapu, Knowledge-based framework for intelligent emotion recognition in spontaneous speech. Proc. Comput. Sci. **96**, 587–596 (2016); Knowledge-based and intelligent information; engineering systems, in *Proceedings of the 20th International Conference KES-2016* (2016)

16. R. Chakraborty, M. Pandharipande, S.K. Kopparapu, Mining call center conversations exhibiting similar affective states, in *PACLIC 2016* (2016)

17. R. Chakraborty, M. Pandharipande, S.K. Kopparapu, Spontaneous speech emotion recognition using prior knowledge, in *23rd International Conference on Pattern Recognition, ICPR 2016, Cancún, Mexico, 4–8 Dec* (2016), pp. 2866–2871

18. R. Chakraborty, M. Pandharipande, S.K. Kopparapu, *Do You Mean What You Say? Recognizing Emotions in Spontaneous Speech* (Springer Singapore, Singapore, 2017), pp. 55–63

19. R. Chakraborty, M. Pandharipande, S.K. Kopparapu, Emotion Datasets (2017), https://sites.google.com/view/emotion-spontaneous-speech/related-information/datasets

20. K. Conway, K.H. Capers, C. Danson, D. Brown, D. Gustafson, R. Warford, M. Moore, Method and system for analyzing separated voice data of a telephonic communication between a customer and a contact center by applying a psychological behavioral model thereto, 26 Nov 2013. U.S. Patent No. 8,594,285 (2013)

21. R. Cowie, E. Douglas-Cowie, N. Tsapatsoulis, G. Votsis, S. Kollias, W. Fellenz, J.G. Taylor, Emotion recognition in human-computer interaction. IEEE Signal Process. Mag. **18**(1), 32–80 (2001)

22. D. Neiberg, K. Elenius, I. Karlsson, K. Laskowski, Emotion recognition in spontaneous speech. Work. Pap. Linguist. **52**, 101–104 (2009)

23. D. Dimitriadis, M.E. Gilbert, T. Mishra, H.J. Schroeter, Real-time emotion tracking system, 31 May 2016. U.S. Patent No. 9,355,650 (2013)

24. D. Ververidis, C. Kotropoulos, A review of emotional speech databases, in *Proceedings of the Panhellenic Conference on Informatics (PCI)* (2003), pp. 560–574

25. E. Douglas-Cowie, R. Cowie, I. Sneddon, C. Cox, O. Lowry, M. Mcrorie, J.-C. Martin, L. Devillers, S. Abrilian, A. Batliner, N. Amir, K. Karpouzis, The humaine database: Addressing the collection and annotation of naturalistic and induced emotional data, in *Proceedings of the 2nd International Conference on Affective Computing and Intelligent Interaction*, ACII '07 (Springer, Berlin, Heidelberg, 2007), pp. 488–500

26. D.-C. Ellen, C. Nick, C. Roddy, R. Peter, Emotional speech: towards a new generation of databases. Speech Commun. **40**, 33–60 (2003)

27. Emo-DB, Berlin Database of Emotional Speech (2013), http://www.emodb.bilderbar.info/

28. I.S. Engberg, A.V. Hansen, O. Andersen, P. Dalsgaard, Design, recording and verification of a danish emotional speech database, in *Fifth European Conference on Speech Communication and Technology* (1997)

29. F. Burkhardt, A. Paeschke, M. Rolfes, W.F. Sendlmeier, B. Weiss, A database of german emotional speech, in *Interspeech*, vol. 5 (2005), pp. 1517–1520

30. Filtering and Noise Adding Tool (2015)

31. B.A. Forouzan, *Data Communications and Networking*, 3rd edn. (McGraw-Hill Inc, New York, 2003)

32. Groningen, Elra-s0020 (1996), http://catalog.elra.info/product_info.php?products_id=61

33. P. Gupta, N. Rajput, Two-stream emotion recognition for call center monitoring, in *INTERSPEECH* (2007)

34. S. Haykin, *Neural Networks: A Comprehensive Foundation*, 2nd edn. (Prentice Hall PTR, Upper Saddle River, 1998)

35. L. He, M. Lech, N. Maddage, S. Memon, N. Allen, Emotion recognition in spontaneous speech within work and family environments, in *International Conference on Bioinfomatics and Biomedical Engineering* (2009), pp. 1–4

36. J.H. Hansen, S.E. Bou-Ghazale, R. Sarikaya, B. Pellom, Getting started with susas: a speech under simulated and actual stress database, in *Eurospeech*, vol. 97 (1997), pp. 1743–46
37. B.-H. Juang, *"Lawrence R Rabiner," Automatic Speech Recognition—A Brief History of the Technology Development*, vol. 1 (Georgia Institute of Technology, Atlanta Rutgers University and the University of California, Santa Barbara, 2005)
38. Y. Kato, T. Kamai, Y. Nakatoh, Y. Hirose, Emotion recognition apparatus, 19 June 2012. U.S. Patent No. 8,204,747 (2012)
39. Y. Khan, C. Huybregts, J. Kim, T.C. Butcher, Real-time emotion recognition from audio signals, 21 Jan 2016. U.S. Patent App. No. 14/336,847 (2016)
40. H.G. Kim, I.J. Kim, J.H. Chang, K.H. LEE, C.S. Bae, Method for emotion recognition based on minimum classification error, 26 Aug 2010. U.S. Patent App. No. 12/711,030 (2010)
41. A. Konchitsky, System and method for detection of emotion in telecommunications, 16 Aug 2007. U.S. Patent App. No. 11/675,207 (2007)
42. S.G. Koolagudi, K.S. Rao, Emotion recognition from speech using source, system, and prosodic features. Int. J. Speech Technol. **15**(2), 265–289 (2012)
43. S.K. Kopparapu, *Non-Linguistic Analysis of Call Center Conversations* (Springer, India, 2015)
44. A. Krishnan, M. Fernandez, System and method for recognizing emotional state from a speech signal, 1 Dec 2011. U.S. Patent App. No. 13/092,792 (2011)
45. L.I. Kuncheva, *Combining Pattern Classifiers: Methods and Algorithms* (Wiley, New York, 2004)
46. R. Laperdon, M. Wasserblat, T. Ashkenazi, I.D. David, O. Pereg, Method and apparatus for real time emotion detection in audio interactions, 28 July 2015. U.S. Patent No. 9,093,081 (2015)
47. D. Le, E.M. Provost, Emotion recognition from spontaneous speech using hidden markov models with deep belief networks, in *2013 IEEE Workshop on Automatic Speech Recognition and Understanding* (2013), pp. 216–221
48. C.M. Lee, S.S. Narayanan, Towards detecting emotions in spoken dialogs. IEEE TASP **13**, 293–303 (2005)
49. A. Liberman, Apparatus and methods for detecting emotions, 28 Oct 2003. U.S. Patent No. 6,638,217 (2003)
50. LibSVM
51. LibSVM (2015), https://www.csie.ntu.edu.tw/~cjlin/libsvm/. Accessed Dec 2015
52. B. Liu, Sentiment analysis and opinion mining (2012)
53. I. Luengo, E. Navas, I. Hernáez, J. Sánchez, Automatic emotion recognition using prosodic parameters, in *INTERSPEECH 2005—Eurospeech, 9th European Conference on Speech Communication and Technology, Lisbon, Portugal, 4–8 Sept* (2005), pp. 493–496
54. K.A.T.C. Marasek, Method for detecting emotions in speech, involving linguistic correlation information, 2 Apr 2003. E.P. Patent App. No. EP20,010,123,079 (2003)
55. K.S.I. Marasek, R.S.I.G.H. Santos, R.S.I.T. GmbH, Method of determining phonemes in spoken utterances suitable for recognizing emotions using voice quality features, 25 Feb 2004. E.P. Patent App. No. EP20,020,018,386 (2004)
56. O. Martin, I. Kotsia, B. Macq, I. Pitas, The enterface05 audio-visual emotion database, in *Proceedings of the 22nd International Conference on Data Engineering Workshops* (IEEE, 2006), pp. 8–8
57. M.L. McHugh, Interrater reliability: the kappa statistic. Biochemia Med. **3**, 276–282 (2012)
58. G. McKeown, M. Valstar, R. Cowie, M. Pantic, M. Schroder, The semaine database: annotated multimodal records of emotionally colored conversations between a person and a limited agent. IEEE Trans. Affect. Comput. **3**, 5–17 (2012)
59. A. Meidan, Method and apparatus for indicating the emotional state of a person, 27 July 1995. W.O. Patent App. No. PCT/US1995/000,493 (1995)
60. T. Mishra, D. Dimitriadis, Incremental emotion recognition, in *INTERSPEECH* (2013)
61. S. Mitsuyoshi, Emotion recognizing method, sensibility creating method, device, and software, 4 Mar 2008. U.S. Patent No. 7,340,393 (2008)

62. T.K. Moon, *Error Correction Coding: Mathematical Methods and Algorithms* (Wiley, New York, 2005)

63. E. Mower, M.J. Mataric, S. Narayanan, A framework for automatic human emotion classification using emotion profiles. IEEE Trans. Audio Speech Lang. Process. **19**(5), 1057–1070 (2011)

64. R. Nakatsu, J. Nicholson, N. Tosa, Emotion recognition and its application to computer agents with spontaneous interactive capabilities, in *Proceedings of the Seventh ACM International Conference on Multimedia (Part 1)*, MULTIMEDIA '99 (ACM, New York, USA, 1999), pp. 343–351

65. M.L. Narayana, S.K. Kopparapu, Effect of noise-in-speech on mfcc parameters, in *Proceedings of the 9th WSEAS International Conference on Signal, Speech and Image Processing, and 9th WSEAS International Conference on Multimedia, Internet & Video Technologies*, SSIP '09/MIV'09, (World Scientific and Engineering Academy and Society (WSEAS), Stevens Point, Wisconsin, USA, 2009), pp. 39–43

66. M.A. Nicolaou, H. Gunes, M. Pantic, Output-associative rvm regression for dimensional and continuous emotion prediction. FG, 16–23 (2011)

67. T. Nwe, S. Foo, L. De Silva, Speech emotion recognition using hidden markov models. Speech Commun. **41**(4), 603–623 (2003)

68. Details of Emotional Databases (2017), https://sites.google.com/view/emotion-spontaneous-speech/related-information/datasets

69. K. Okim, D.O. Johnson, Systems and methods for automated evaluation of human speech, 1 Sept 2016. W.O. Patent App. No. PCT/US2016/019,939 (2016)

70. openSMILE (2015), http://www.audeering.com/research/opensmile

71. P.Y. Oudeyer, The production and recognition of emotions in speech: features and algorithms. Int. J. Hum. Comput. Stud. **59**, 157–183 (2003)

72. P.Y. Oudeyer, Emotion recognition method and device, 11 Nov 2008. U.S. Patent No. 7,451,079 (2008)

73. M.A. Pandharipande, S.K. Kopparapu, Audio segmentation based approach for improved emotion recognition, in *TENCON 2015–2015 IEEE Region 10 Conference* (2015), pp. 1–4

74. M.A. Pandharipande, S.K. Kopparapu, A novel approach to identify problematic call center conversations, in *2012 International Joint Conference on Computer Science and Software Engineering (JCSSE)* (2012), pp. 1–5

75. M. Pandharipande, R. Chakraborty, S.K. Kopparapu, Methods and challenges for creating an emotional audio-visual database, in *Oriental COCOSDA* (South Korea, Seoul, 2017), p. 2017

76. M. Pandharipande, S.K. Kopparapu, Audio segmentation based approach for improved emotion recognition, in *TENCON*, (Macua, China, 2015)

77. B. Pang, L. Lee, Opinion mining and sentiment analysis. Found. Trends Inf. Retr. **2**(1–2), 1–135 (2008)

78. D. Pappas, I. Androutsopoulos, H. Papageorgiou, Anger detection in call center dialogues, in *2015 6th IEEE International Conference on Cognitive Infocommunications (CogInfoCom)* (2015), pp. 139–144

79. D. Pappas, I. Androutsopoulos, H. Papageorgiou, Anger detection in call center dialogues, in *2015 6th IEEE International Conference on Cognitive Infocommunications (CogInfoCom)* (2015), pp. 139–144

80. O. Pereg, M. Waserblat, Apparatus and methods for the detection of emotions in audio interactions, 15 Feb 2007. W.O. Patent App. No. PCT/IL2005/000,848 (2007)

81. O. Pereg, M. Wasserblat, Apparatus and methods for the detection of emotions in audio interactions, 14 Feb 2008. U.S. Patent App. No. 11/568,048 (2008)

82. V. Petrushin. Emotion in speech: Recognition and application to call centers. In *Artificial Neural Networks in Engineering (ANNIE)*, pages 7–10, 1999

83. V.A. Petrushin. A system, method, and article of manufacture for detecting emotion in voice signals through analysis of a plurality of voice signal parameters, 8 Mar 2001. C.A. Patent App. No. CA 2,353,688

84. V.A. Petrushin. Detecting emotions using voice signal analysis, 1 Dec 2009. U.S. Patent No. 7,627,475

85. V.A. Petrushin, Detecting emotion in voice signals in a call center, 10 May 2011. U.S. Patent No. 7,940,914 (2011)

86. V.A. Petrushin, Emotion in speech: Recognition and application to call centers, in *In Engr* (1999), pp. 7–10

87. S. Planet, I. Iriondo, Improving spontaneous children's emotion recognition by acoustic feature selection and feature-level fusion of acoustic and linguistic parameters, in *NOLISP* (2011), pp. 88–95

88. R. Plutchik, *The Psychology and Biology of Emotion* (HarperCollins College Publishers, 1994)

89. J. Pohjalainen, F. Fabien Ringeval, Z. Zhang, B. Schuller, Spectral and cepstral audio noise reduction techniques in speech emotion recognition, in *Proceedings of the 2016 ACM on Multimedia Conference*, MM '16 (2016), pp. 670–674

90. T. Polzehl, A. Schitt, F. Metze, *Spoken Dialogue Systems Technology and Design, chapter Salient Features for Anger Recognition in German and English IVR Portals* (Springer, New York, 2011)

91. T. Polzehl, A. Schmitt, F. Metze, M. Wagner, Anger recognition in speech using acoustic and linguistic cues. Speech Commun. **53**(9), 1198–1209 (2011); Sensing emotion and affect—facing realism in speech processing

92. L.R. Rabiner, B.H. Juang, *Fundamentals of Speech Recognition* (Prentice-Hall, Englewood Cliff, New Jersey, 1993)

93. F. Ringeval, A. Sonderegger, J. Sauer, D. Lalanne, Introducing the recola multimodal corpus of remote collaborative and affective interactions, in *2013 10th IEEE International Conference and Workshops on Automatic Face and Gesture Recognition (FG)* (2013), pp. 1–8

94. D. Bernard Ryan (2003), http://www.denisryan.com/thesis/capstoneV1.02.pdf

95. B. Schuller, A. Batliner, S. Steidl, D. Seppi, Recognising realistic emotions and affect in speech: State of the art and lessons learnt from the first challenge. Speech Commun. **53**, 1062–1087 (2011)

96. B. Schuller, S. Reiter, R. Muller, M. Al-Hames, M. Lang, G. Rigoll, Speaker independent speech emotion recognition by ensemble classification, in *ICME* (2005), pp. 864–867

97. B. Schuller, S. Steidl, A. Batliner, The interspeech 2009 emotion challenge, in *INTERSPEECH* (2009), pp. 312–315

98. B.W. Schuller, D. Seppi, A. Batliner, A.K. Maier, S. Steidl, Towards more reality in the recognition of emotional speech, in *ICASSP* (2007), pp. 941–944

99. B.W. Schuller, S. Steidl, A. Batliner, The interspeech 2009 emotion challenge, in *INTERSPEECH* (2009), pp. 312–315

100. B. Schuller, D. Arsi, F. Wallhoff, G. Rigoll, Emotion recognition in the noise applying large acoustic feature sets, in *Speech Prosody* (2006)

101. B. Schuller, S. Steidl, A. Batliner, A. Vinciarelli, K. Scherer, F. Ringeval, M. Chetouani, F. Weninger, F. Eyben, E. Marchi, M. Mortillaro, H. Salamin, A. Polychroniou, F. Valente, S. Kim, The interspeech 2013 computational paralinguistics challenge: social signals, conflict, emotion, autism, in *Proceedings of the Annual Conference of the International Speech Communication Association, INTERSPEECH* (2013), pp. 148–152, 08

102. M. Sjöberg, Y. Baveye, H. Wang, Vu Lam Quang, B. Ionescu, E. Dellandréa, M. Schedl, C.-H. Demarty, L. Chen, The mediaeval 2015 affective impact of movies task, in *MediaEval* (2015)

103. M. Sjöberg, Y. Baveye, H. Wang, V. Lam Quang, B. Ionescu, E. Dellandréa, M. Schedl, C.-H. Demarty, L. Chen, The mediaeval 2015 affective impact of movies task, in *MediaEval 2015 Workshop* (2015)

104. I. Sneddon, M. McRorie, G. McKeown, J. Hanratty, The belfast induced natural emotion database. IEEE Trans. Affect. Comput. **3**(1), 32–41 (2012)

105. R. Srinivasan, *Recognition of emotion from speech: A review Modeling and Recognition-Algorithms and Applications* (In Speech Enhancement, InTech, 2012)

106. A. Tarasov, S.J. Delany, Benchmarking classification models for emotion recognition in natural speech: a multi-corporal study, in *FG* (2011), pp. 474–477
107. R. Tato, T. Kemp, K. Marasek, Method for detecting emotions involving subspace specialists. U.S. Patent No. 7,729,914 (2010)
108. WEKA Toolkit (2015), http://www.cs.waikato.ac.nz/ml/weka/
109. K.P. Truong, S. Raaijmakers, Automatic recognition of spontaneous emotions in speech using acoustic and lexical features, in *Machine Learning for Multimodal Interaction: 5th International Workshop, MLMI 2008*, Utrecht, The Netherlands, 8–10 Sept 2008, ed. by A. Popescu-Belis, R. Stiefelhagen (Springer, Berlin, Heidelberg, 2008), pp. 161–172
110. M. Valstar, B. Schuller, K. Smith, F. Eyben, B. Jiang, S. Bilakhia, S. Schnieder, R. Cowie, M. Pantic, Avec 2013: the continuous audio/visual emotion and depression recognition challenge, in *Proceedings of the 3rd ACM International Workshop on Audio/visual Emotion Challenge* (2013), pp. 3–10
111. L. Vidrascu, L. Devillers, Five emotion classes detection in real-world call center data the use of various types of paralinguistic features, in *PARALING* (2007), pp. 11–16
112. A.J. Viera, J.M. Garrett, Understanding interobserver agreement: the kappa statistic. Fam. Med. **37**(5), 360–363 (2005)
113. T. Vogt, E. Andre, Comparing feature sets for acted and spontaneous speech in view of automatic emotion recognition, in *ICME* (2005), pp. 474–477
114. S. Wu, T.H. Falk, W.Y. Chan, Automatic speech emotion recognition using modulation spectral features. Speech Commun. **53**(5), 768–785 (2010)
115. Xie, Z., Guan, L.: Multimodal information fusion of audiovisual emotion recognition using novel information theoretic tools, in *2013 IEEE International Conference on Multimedia and Expo (ICME)* (2013), pp. 1–6
116. S. Yacoub, S. Simske, X. Lin, J. Burns, Recognition of emotions in interactive voice response systems, in *Proceedings of Eurospeech* (2003), pp. 729–732
117. Z. Zeng, M. Pantic, G.I. Roisman, T.S. Huang, A survey of affect recognition methods: audio, visual, and spontaneous expressions. IEEE Trans. Pattern Anal. Mach. Intell. **31**(1), 39–58 (2009)

Index

© Springer Nature Singapore Pte Ltd. 2017
R. Chakraborty et al., *Analyzing Emotion in Spontaneous Speech*,
https://doi.org/10.1007/978-981-10-7674-9

Printed in the United States
by Bookmasters

Printed in the United States
By Bookmasters